Technical Change
and Full Employment

edited by

Christopher Freeman and Luc Soete

Basil Blackwell

First published 1987

Basil Blackwell Ltd
108 Cowley Road, Oxford, OX4 1JF, UK

Basil Blackwell Inc.
432 Park Avenue South, Suite 1503
New York, NY 10016, USA

British Library Cataloguing in Publication Data
Technical change and full employment.
    1. Labor supply – Great Britain –
    Effect of technological innovations on
    I. Freeman, Christopher   II. Soete, Luc
    331.12′5   HD6331.2.G7

    ISBN 0-631-14099-9

Library of Congress Cataloging in Publication Data
Technical change and full employment.
    Bibliography: p.
    Includes index.
    1. Labor supply – Great Britain – Effect of
technological innovations on.   I. Freeman, Christopher.
II. Soete, Luc.
HD6331.2.G7T39   1987   331.12′5′0941   87-6637
ISBN 0-631-14099-9

Typeset in 10 on 12pt Times
by Dobbie Typesetting Service, Plymouth, Devon
Printed in Great Britain by T. J. Press (Padstow) Limited

Contents

Acknowledgments

This book presents the results of a research programme on Technical Change and Employment Opportunities (TEMPO) which was carried out by a small team at the Science Policy Research Unit (SPRU) at the University of Sussex.

For the main part the research was supported by a programme grant from the Economic and Social Research Council (ESRC), with supporting grants from IBM for computer modelling work (described in chapter 6) and for a review of comparable work in other industrialized countries. Their support is gratefully acknowledged.

In the first year of the programme the team was led by Charles Cooper. We are particularly grateful to him for his initial impetus. He was also reponsible for the first book to emerge from the programme – a review of the economics literature relating to technical change and employment (Cooper and Clark, 1982). Our own original theoretical framework was also largely inspired by this early work, in so far as Cooper and Clark's survey pointed to the need for a different approach. This new approach was described in a second book (Freeman, Clark and Soete, 1982) parts of which are summarized and further developed in chapters 2–4 of this book.

In parallel with this work, Marie Jahoda, who pioneered the study of the psychological effects of unemployment in the great Depression before the Second World War, resumed this interest in the contemporary situation. This led to her book on *Employment and Unemployment: A Social-psychological Analysis* (Jahoda, 1982) which is the basis of the first chapter in this book. We are extremely grateful to her, not just for her present contribution, but for her full participation in all the work of the research team, and especially in improving all the many discussion drafts of papers and publications.

In addition to Charles Cooper and Marie Jahoda, the support of all the other members of the TEMPO team who contributed to our work at various stages of the programme is gratefully acknowledged. In particular, we would like to thank John Clark, who developed the computer simulation model used as a research tool throughout the programme and continued to contribute to our work on a part-time basis after he left the team in 1982. Following his departure, Pari Patel has helped to fill the gap and together with John Clark has contributed to many aspects of the team's work, as shown in this book. We are deeply indebted to them and to Roy Turner for their numerous contributions. In developing our critique of other mathematical models which are widely used in the United Kingdom, we were also fortunate to have the benefits of fruitful discussions with members of the University Economics faculty, particularly Rob Eastwood who wrote chapter 5 of this book.

The main thrust of the research, following the development of the theoretical framework and simulation model, was a series of detailed studies of all the major sectors of UK manufacturing and service industries. We are particularly grateful to Ken Guy for his contribution to these studies, as well as for much other work in relation to the programme as a whole.

For each of the sectors which were the subject of detailed study, a preliminary background paper was prepared. In most cases this was carried out by members of the TEMPO team, but in a few cases by external consultants with expertise in particular sectors, or by other members of SPRU with specialized expertise or by postgraduate students. In the preparation of the papers particular attention was paid to the collection and analysis of information relating to the trends of technical innovation and invention. The sources used included the SPRU data banks on patents and innovations, which now cover all sectors of the economy.

Keith Pavitt, Joe Townsend and Mike Robson provided essential help in this work. For the analysis of the main trends in investment, output, productivity, and foreign trade we initially made use of the Cambridge–Warwick growth model data as well as relevant work by economists on the specific sector under investigation and information from firms. We are grateful to all our colleagues inside and outside SPRU and to the Warwick Institute for Employment Research for their help with these data, and with these sector papers.

The sector papers were circulated in each case to a small group of experts, who were usually invited to attend a 'brainstorming' one-day workshop; these were used to consolidate and check information contained in the corresponding background papers and for discussion of hypotheses put forward. After the workshop a final report was then drawn up for each sector based on the draft background paper, the workshop discussions, subsequent correspondence and interviews, and more recent statistics. These final reports were published in five volumes by Gower Press, and are listed in the References. We are grateful to all authors of the sector studies listed there.

The experience of the workshops was encouraging, both in terms of participation from industry, government and academia and in terms of relevant comment, criticisms, suggestions and discussion. At most workshops major industrial firms and industrial organizations as well as representatives of NEDO and government departments have been present and contributed substantially to the discussion. There was also a good response in terms of written comments and additional information after each workshop. We are grateful to all the participants for this constructive involvement.

In this book we attempt to summarize and synthesize the results of these sector studies in chapters 8–10.

All of this research and publication involved a significant amount of typing and revision of successive drafts of books and papers. It would have been quite impossible to get through it without the organizing skills and tremendous efforts of Linda Gardiner, the Secretary to the Programme. We are particularly grateful to her for her contribution.

Last but not least we would like to thank Glynis Flood for bringing this book to a good end. She won our admiration for the indefatigable good spirits and even humour with which she reprocessed draft after draft, which we never stopped producing even in EX-TEMPO time and at EX-TEMPO speed but which we always received back in IN-TEMPO.

Christopher Freeman
Luc Soete

Introduction

Christopher Freeman and Luc Soete

When the research programme 'Technical Change and Employment Opportunities', upon which this book reports, was started in 1979, unemployment in the United Kingdom stood at 1.4 million or some 5.3 per cent of the total civilian labour force. Today, with unemployment – on a similar definition – between two and three times higher, such unemployment rates would be readily identified with 'full' employment. Yet the original TEMPO research proposal originated from a feeling that the steady, but slow, growth in unemployment over the seventies in the United Kingdom and many European countries was already indicative of an emerging structural unemployment problem, which, with the further diffusion of 'automation' technologies and increased international competition, might well worsen over the eighties.

Indeed, at that time we explicitly rejected (Freeman, 1979; Soete, 1978) the view that the rise in unemployment over the seventies was of the conventional 'cyclical' type, exacerbated by an unfortunate external oil-shock crisis, to which labour markets would adjust in a relatively straightforward way, a view that was prevalent in many quarters (see, e.g., the OECD's McCracken Report, 1977). For us, the emerging employment crisis, already quite obvious in some of the smaller European countries such as Holland and Belgium, could be related directly to structural problems as well as to some of the more strongly emerging rationalization and automation pressures, reflected in the 'labour-saving' nature of most investment outlays and increased international competition, and the inevitable lags and imperfections in the employment compensation mechanisms. These arguments were summarized in a book written in 1980–81 and published in 1982 under the title *Unemployment and Technological Innovation* (Freeman et al., 1982).

Five years later, after having witnessed the emergence and persistence of large-scale unemployment in the United Kingdom and most other

1

European countries, we present the results of our research programme in this book under the title *Technical Change and Full Employment*. This is not just a fashionable change in title. It reflects a significant turn in what, in our view, the focus of analysis in this area should be.

The question whether the rise in unemployment over the seventies and eighties was primarily related to technical change or not remains undoubtedly at the forefront of our analysis. We continue to reject those economic analyses which either completely ignore technical change, such as many contributions in the field of labour economics, or those which simply reiterate the general proposition of compensation theory without any regard to the specific circumstances within which this mechanism is supposed to be operating, or without any consideration of the time-lags inevitably involved. Such 'dogmatic' approaches will, in the present context of mass unemployment and the emergence of some radically new technologies, continue, if anything, to reinforce the very fears they were intended to allay.

But rather than simply restating these arguments, we aim in this book to illustrate the way in which technical change can bring about employment growth and how the process of 'creative' capital and skill destruction can generate the necessary conditions for long-term economic recovery. We focus in particular on those adjustment policies which can reinforce the employment compensation mechanism and reduce the time-lags involved in the adjustment of the capital stock and the skill profile of the labour force. If technical innovation is to be readily and widely promoted by the public at large, rather than lukewarmly accepted or sullenly resisted, this can only be on the basis of a deeper and more realistic understanding of the potential of these new technologies for job creation, as well as the complexity of the process of transition during which such new technologies are diffused and assimilated and their job-displacing effects are absorbed. Such an understanding is also essential if economic, industrial and social policies are to be adopted which will help, rather than hinder, this transition. Economic theory, although always recognizing the critical importance of technical innovations and their diffusion for the dynamism of any economy, has traditionally kept rather aloof from the specifics of this process. The purpose of this book is a modest attempt to compensate for this traditional neglect, and to bring out the importance of policies for technology as an integral component of economic and social policies for adjustment and recovery.

Indeed, it is crucial that the policy implications of our analysis are increasingly recognized as the vital technology component of any macroeconomic employment recovery policy. Our previous emphasis (Freeman et al., 1982, chapter 10) on innovation policy and broadly Keynesian, but technology-related, infrastructural investment remains, and is re-emphasized in chapter 12 of this book. However, we stress two additional points. First, many of the technology supply side policies suggested in our previous work (some of which have been introduced in the United Kingdom and other European countries) would probably have only limited growth and employment effects so long as there is a problem of inadequate aggregate demand. It could be said that even if these policies were introduced today in their totality, one would only be witnessing the direct employment effects of such policies – effects which are by and large limited to the 'high-tech' industries and will be relatively small. The more important indirect employment effects through the use and more widespread application of new technology will barely surface. Only with aggregate demand sufficiently present could one witness the various technology-related output multiplier and employment generating compensation effects. In other words, despite our emphasis on the structural nature of the present employment crisis and the need for structural, technology-related adjustment policies, we come down completely on the side of those who would claim that insufficiency of demand in the UK and Europe at large is one of the crucial bottlenecks in preventing a return to full employment. But unlike most neo-Keynesian analyses we would continue to claim that taken on its own demand 'reflationary' policies would be inadequate and likely to fail if not accompanied or, more precisely, if not fully integrated with the sort of structural technology policies set out in chapter 12.

Secondly, the present employment crisis is also a reflection of an increased need for institutional change. The emergence and efficient assimilation of 'major new technological systems' (Freeman et al., 1982), such as electric power, the internal combustion engine, in previous times, or microelectronics today, involves a prolonged process of social adaptation. Characteristically, technological change proceeds much more rapidly than social change. Institutions and attitudes which were once favourable to earlier forms of technical and social development may now become an impediment to the spread of new technologies. Generally speaking there is often a high degree of inertia in the social and institutional framework. Following Perez (1983), we would argue that we are now confronted with a severe 'mismatch' between the emerging new technologies and the social framework, resulting both in sluggish investment behaviour and structural unemployment. The capacity to adapt the social–institutional framework to meet the new requirements for the emerging new technology, and especially to develop the new skills required at all levels in the labour force, has however in recent times not been particularly high in the United Kingdom with major repercussions on its international competitive performance; and historically a long-term loss of technological leadership. Indeed, by no means is the road to full employment an easy one.

We discuss these issues at greater length in the final chapters on policy (part IV). We start this book, though, on the basis of the analysis of Marie Jahoda (chapter 1), from the standpoint that the prevalence of large-scale unemployment, as in the United Kingdom and many other European countries, is a dangerous social evil, as well as a massive waste of resources in purely economic terms. It is important to restate this proposition in unambiguous form, because whilst it would have been almost a platitude during the period of 'full employment' in the fifties and sixties its validity was increasingly contested in the seventies and early eighties, and the priority accorded to employment policies was progressively undermined. More recently (in 1984–6) there have been signs of a return to the priority which was almost universally accepted in the quarter century after the Second World War.

However, even amongst those who would still accord a high social and economic priority to the goal of full employment there are many who doubt the feasibility of attaining this goal. Some doubt this because of problems of inflation or because of belief in some 'natural' level of unemployment which governments are supposedly powerless to change. Others believe that the goal of 'full employment' is now no longer attainable, because of the nature of contemporary technical change.

This book questions those beliefs. As before, our concern is with the analysis of the influence of technical change on past, present and future levels of employment in the United Kingdom. We certainly would accept that the short-term unemployment outlook is still bleak at the time of writing (1985–86) even taking into account some growth in service employment. Indeed our own analysis in chapters 7 and 8 points to some factors which may exacerbate the present problems.

But we shall argue that in the medium and long term these trends can be reversed; that technical change and full employment are indeed fully complementary; and that new technologies could play a very important part in generating new employment opportunities, as well as regenerating older industries and services. This argument is based in part on historical analogy, in part on theoretical analysis, in part on cross-country comparisons and in part on the type of empirical evidence about new areas of employment growth discussed in chapters 8, 9 and 10.

As before the analysis presented here does indeed show that over the last 10 years there has been a great deal of labour-displacing technical change in most industries and that there is likely to be a great deal more in the future. It is essential to remember, though, that this was also true of the period of 'full employment' in the fifties and sixties. In fact, the rate of growth of labour productivity was at that time rather higher than in the decade which followed. The main difference was that new employment opportunities were being generated at that time at a rate which outstripped both the loss of old jobs and the increase in the labour force associated with demographic and social trends.

In discussing the influence of technology on employment and unemployment it is essential, therefore, to take into account not merely the labour-saving trends in technology and the loss of jobs which this may entail, but also the creation of entirely new opportunities for employment in the production and delivery of new products and services, as well as the 'compensating' effect of growth in demand for the existing range of output, which is made possible by the productivity, price, demand and income effects of technical change.

Historically speaking, over the past two centuries it is incontestable that the combined effect of these forces, together with the persistent reduction in working hours, has made it possible to generate millions of new jobs on a scale which more than offset the tendency to rising unemployment as a result of labour-saving technical change. The questions which must be asked, therefore, are: first, why has this process been periodically disrupted? Secondly, are there special features about the present trends in technology and in the economy which would permanently (or for a long time) prevent a return to another period of high growth in output and in employment, as occurred after previous periods of deep structural change and recession?

In our earlier book (Freeman et al., 1982) we attempted to answer the first of these two questions, arguing that, whilst there are multiple and complex causes of deep recessions, prolonged periods of slack demand for labour have been associated with the temporary exhaustion of the impetus to new investment and employment, derived from one technological 'system' and difficulties in the adoption of a new one. In this book we attempt also to answer the second question in relation to the technological revolution which is now transforming the United Kingdom and the world economy.

This conceptualization of the growth process is associated primarily with Schumpeter (1939) and his notion of 'creative gales of destruction' arising from the introduction of major new technologies into the economy. We do not wish here to repeat the analysis presented in our earlier work. It is essential, however, to reiterate some points in relation to the trends in the UK economy, which are our main concern in this book. It is also essential to indicate some major differences of emphasis which distinguish our approach from that of some other economists and lead us to somewhat different policy conclusions. We deal with these differences in chapters 2, 3 and 4. With chapter 1, these chapters form the theoretical core of the book.

Following this critical clarification of some of the basic issues in the theory of technical change and employment we discuss a possible methodological framework in chapters 5 and 6 and present some model simulation results in chapter 7, grouped in part II. Although we are only too well aware of the limitations of mathematical models of the economic system, we believe that there are some advantages in attempting to simulate, in a formal vintage model, the ways in which new investment and technical change affect levels of employment. Following upon a critique of one of the established models of the UK economy in chapter 5, we go on to describe the simulation model which we used in our own work in chapter 6. This model is mainly applicable in those sectors of the economy where much technical change is embodied in new capital development. Thus we use it in chapter 7, in relation to the UK manufacturing sector and its future employment growth potential.

The main sectoral results of our research programme are presented in part III. In chapter 8 we present the results for the various UK manufacturing subsectors, in chapter 9 for the service sectors and in chapter 10 for the 'new' emerging information technology sector. Readers are referred to the volumes listed on pages 274–275 for details of the sector studies. In this book we synthesize the overall results without going into a great deal of detail about any one sector.

Part IV contains the two concluding chapters. Chaper 11 discusses the inter-relationships between the various sectors and presents some employment forecasts. Chapter 12 points to some policy implications.

I
THEORY

Like most studies in the field, our analysis concentrates on the economic aspects of the problem of technical change and employment. The problem of job-generating technical change is by no means a new one for economists and it has been analysed and debated by them for over two centuries. Their analysis, summarized in chapter 2, is the essential starting point for any serious discussion of the problem. However, this should not be taken as reflecting any lack of concern with the wider social dimensions and still less with the technological aspects. On the contrary, as will become apparent, one of our main 'leitmotivs' is that a purely economic analysis, in the narrow sense of the term, is not adequate in itself.

We consequently start this theoretical part with a wider, socio-psychological look at the present unemployment problem and conclude it with a detailed look at the technological aspects of the major 'cluster' of new technologies affecting employment today: information technology. Few people would be better placed than Marie Jahoda to tackle the first issue. She was one of the first to study the destructive social psychological consequences of prolonged unemployment in Marienthal, a village in Austria, in the thirties. Her unique comparative analysis with the eighties situation in the United Kingdom brings out the limitations of the social welfare system in coping with the harmful effects of long-term unemployment. Unemployment today with its long-term predominance is, if anything, in its socio-psychological consequences more harmful to the individual than in the thirties. Jahoda's analysis questions therefore directly those schools of economic thought which put the main emphasis today on 'voluntary' unemployment, and see the problem primarily as one of reducing benefit levels to provide greater 'incentives' in a supposed trade-off between work and leisure.

The review of the economics literature in chapter 2 does not enter this debate, rather it sticks to the narrower issue of the employment implications of technical change. It is necessarily brief and does not claim to be complete. Paradoxically though, despite the voluminous literature on the subject, the chapter brings out the present-day neglect, if not total disinterest, on the part of the academic economics profession in the issue. Part of this neglect is, in our view, a reflection of a more widespread general neglect of the technical change factor in macroeconomics and the focus on general equilibrium as the main approach to the understanding of the economy. This focus detracts from concern with the actual process of technical change and leads to an abiding preoccupation with factor prices and market mechanisms. If anything, there is in our view an essential need to go back to the way classical economics (and Schumpeter) focused on the issue of new technologies. This brings to the forefront two essential characteristics which are dealt with in chapters 3 and 4.

First, as discussed at length in chapter 3, nowhere is the gap between economic theory and technical reality further apart than in relation to the representation of the technology factor in the production function. This has been acknowledged by many economists over the last decades, but seems to have been ignored more often than not when dealing with the employment implications of technical change. The 'bias' in the direction of technical change and its relationship with technological trajectories and long-term trends in factor prices is at the core of the time-lags and imperfections in employment compensation. In so far as new technologies will be 'embodied' in new capital formation, the 'classical' assumption of 'fixed' coefficients appears also closer to reality. The non-malleability of capital becomes consequently an essential theoretical ingredient of the methodology developed in part II.

Second, and enlarging this issue to some of the more specific technological aspects of information technology, chapter 4 illustrates in detail the radical nature of the new technology. The emergence during the seventies of a cluster of inter-related new technologies centred around microelectronics is 'revolutionary' in its potential growth and employment implications. We argue that it corresponds to the emergence of a new techno-economic paradigm, 'sudden' in its emergence, 'destructive' in its dislocation and obsolescence effects on skills and capital, 'creative' in its need for radical new skills and capital and its capacity to generate new products and services. Again, it could be said that the fear that the employment displacement effects of these new technologies will predominate over the compensating job creation effects is reminiscent of the classical economic debate. As in the latter case it also is a reflection of the times: a set of 'revolutionary' new technologies and the persistence of high unemployment.

1
Unemployment: Facts, Experience and Social Consequences

Marie Jahoda

The slowdown in economic growth in the seventies and eighties hit the entire Western industrialized world, and Britain particularly hard. Even though this 'first' world is still richer than the second, and very much richer than the Third World, there are now in most industrialized countries millions of people to whom the recession has brought unemployment and with it economic misery.

In Britain estimates of the financial burden imposed on the public purse by the current extent of unemployment vary with political allegiance, but even if one assumes the lowest estimates, it is a staggering amount that no government can face with equanimity. Indeed, no government does. The public debate about unemployment is not about welcoming or rejecting it, but about its causes, consequences and appropriate policies for its reduction, if not elimination.

This debate is necessarily conducted predominantly in economic terms, but there is a growing realization that social and psychological issues are involved which defy translation into the language of economics, although they must be assumed to affect the entire web of society, including its present and future economic performance.

Some economic facts

The OECD estimated that nearly 32 million people in its member countries were unemployed early in 1985; this is double the number reported for 1975 (OECD, 1985). In the United Kingdom more than 3.3 million people were 'officially' registered as unemployed in 1985, and there is so far no

Table 1.1 Unemployment rates (civilian labour force basis) by age, 1980–84

Age group	United States	Canada	Australia	Japan	France[a]	Germany[a]	Great Britain[b]	Italy	Sweden
1980									
All working ages	7.1	7.5	6.1	2.0	6.1	2.7	6.6	3.9	2.0
teenagers[c]	17.8	16.2	17.1	4.2	25.9	3.5	15.7	18.4	7.7
20–24 years	11.5	11.0	8.9	3.3	13.0	3.5	10.3	12.1	3.7
25 years and over	5.1	5.4	3.7	1.8	4.3	2.4	4.9	1.7	1.4
1981									
All working ages	7.6	7.5	5.8	2.2	7.0	3.6	10.1	4.3	2.5
teenagers[c]	19.6	16.2	15.6	5.6	29.1	4.3	21.1	20.9	9.6
20–24 years	12.3	11.2	8.2	3.7	15.1	5.1	15.9	13.0	4.9
25 years and over	5.4	5.6	3.7	2.0	5.0	3.3	7.9	1.9	1.8
1982									
All working ages	9.7	11.0	7.1	2.4	7.8	5.3	11.8	4.8	3.1
teenagers[c]	23.2	21.9	18.5	5.6	31.3	6.9	24.1	23.7	10.9
20–24 years	14.9	16.8	10.4	4.0	17.3	8.0	18.0	14.2	6.0
25 years and over	7.4	8.4	4.7	2.1	5.6	4.8	9.3	2.1	2.3
1983									
All working ages	9.6	11.9	9.9	2.7	8.0	(d)	11.6	5.3	3.5
teenagers[c]	22.4	22.2	23.6	6.4	30.7	(d)	23.4	26.6	10.6
20–24 years	14.5	18.5	14.6	4.1	18.8	(d)	18.2	15.6	7.0
25 years and over	7.5	9.4	6.9	2.4	5.8	(d)	9.1	2.5	2.6
1984									
All working ages	7.5	11.3	9.0	2.8	(d)	(d)	11.6	(d)	3.1
teenagers[c]	18.9	20.0	22.3	6.9	(d)	(d)	22.8	(d)	5.0
20–24 years	11.5	16.8	12.9	4.6	(d)	(d)	18.8	(d)	6.7
25 years and over	5.8	9.3	6.3	2.5	(d)	(d)	9.1	(d)	2.6

[a] French data are for March 1980, 1981, and 1983 and April–May 1982 and German data are for April 1980 and 1982 and May 1981.
[b] Data not adjusted to US concepts. Figures for 1983 are not comparable to the earlier figures because of a change in the system of counting the unemployed from registrations to claimants. The 1983 figures are slightly understated for comparison with earlier years.
[c] 16- to 19-year-olds in the United States, France, Great Britain, and Sweden; 15- to 19-year-olds in Canada, Australia, Japan, and Germany, and 14- to 19-year-olds in Italy.
[d] Not available.

clear sign of a decline in numbers. Here, as in most other OECD countries, unemployment rates are particularly high among those aged under 25 and over 55 years (table 1.1); in the United Kingdom as in the United States the rates are much higher for young members of ethnic minorities than for other groups.

Some suggest that the official statistics overestimate the extent of unemployment because they include fraudulent claimants who are actually working in the 'black' or even the 'white' economy; others, that an underestimate is involved because of the exclusion of discouraged workers who no longer bother to register as unemployed. While it is virtually impossible to arrive at an accurate identification of the size of either group, there has until recently been agreement that notwithstanding their inevitable imperfections the official data are good enough to indicate the seriousness

Table 1.2 Long-term unemployment in 16 OECD countries (as % of total unemployment)

	1979		1984	
	6 months and over	12 months and over	6 months and over	12 months and over
Australia	38.0	18.1	51.3	31.2
Austria	19.4	8.6	28.3	12.9
Belgium	74.9	58.0	81.3	68.0
Canada	15.6	3.5	26.0	9.9
Finland	41.5	19.3	38.6	14.4
France	55.1	30.3	66.5	42.3
Germany	39.9	19.9	55.1	32.7
Ireland	47.9	31.8	57.8	39.1
Italy	n.a.	35.8	n.a.	n.a.
Japan	36.7	16.5	37.6	15.2
Netherlands	49.3	27.1	75.0	55.5
Norway	11.0	3.8	28.5	10.8
Spain	51.6	27.5	73.2	54.2
Sweden	19.6	6.8	27.9	12.3
United Kingdom	40.0	24.8	60.2	39.8
United States	8.8	4.2	19.1	12.3

n.a.: not available.

Note: Measures of long-term unemployment are particularly uncertain, and international comparisons are difficult since the type of source and the definitions used vary from country to country. Data for Australia, Canada, France, Italy, Japan, Norway, Spain, Sweden and the United States are from household surveys and measure the duration of job-search among the unemployed; data for the remaining countries are based on the duration of registration as unemployed according to the records maintained by employment offices. In the case of Canada, France, Italy, Japan (1984), Norway and Spain, persons for whom no duration of unemployment was specified are excluded from total unemployment.

of the problem. With regard to the United Kingdom the various 'official'
changes in the total numbers 'registered' as unemployed over the last
3 years, seem to imply a significant underestimation.

The overall data include, of course, frictional unemployment of short
duration through the normal change of jobs. The average duration of
unemployed periods has, however, risen much faster than frictional
unemployment. According to OECD estimates the long-term unemployed
account now in most European countries for 30 per cent of the total on
a 12-month definition and around 60 per cent or more of the total on a
6-month definition (1983, p. 8). As illustrated in table 1.2, in countries
with a low general unemployment rate this proportion is much lower. In
Austria, for example, only 28.3 per cent of the unemployed had spells
of 6 months or more in 1984; in contrast this percentage was 60 in the
United Kingdom in 1984, to which it had risen from 22.6 per cent in 1962,
a period of relatively full employment. The average duration of a spell
of unemployment in the United Kingdom was 8.9 months in 1982. There
are, of course differences in this respect between various categories of
unemployed: between men and women, age groups and skill groups. 'In
general', the OECD concludes, 'long-term unemployment is now affecting
a higher proportion of men and women in all age groups and skill and
professional categories' (1983, p. 8).

It is, of course, not for the first time in this century that the Western
world has suffered from mass unemployment. During the Great Depression
of the thirties the national rates of unemployment were often even higher
than they are now; in Britain the absolute number of people affected and
the duration is, however, higher now than it was then. Whether a calculus
of deprivation should be based on percentages or absolute numbers is,
perhaps, a debatable point. What is beyond debate is that deprivation is
involved today, as it was then, even though the standard of living, including
that of the currently unemployed, is now considerably higher. During the
thirties the unemployed in several countries lived in abject poverty on a
bare subsistence level and sometimes below. Whether their physical and
psychological suffering, documented in more than 100 studies (summarized
in Eisenberg and Lazarsfeld, 1938), was due only to that poverty or also
to the enforced exclusion from work itself is now hard to disentangle.
Whatever the relative importance of these factors, that situation in Central
Europe was one major cause for Hitler's success.

At present the majority of the unemployed live in relative, not absolute
poverty. Relative deprivation is, of course, still deprivation, all the more
so because the standard of comparison for the unemployed is not the lives
their grandparents lived but their own before they lost their jobs. While
it is not easy to obtain an objective picture of the degree of income
deprivation among the unemployed, several efforts in that direction

have been made. They are important because the notion of 'voluntary' unemployment, based on the assumption that financial deprivation is minimal, plays a considerable role in the public debate.

Kay and Morris (1982) estimated that 2 per cent of the British unemployed were better off living on public support in 1981 than in their previous jobs, and a further 4 per cent received 90 per cent or more of their previous earnings. Given the inevitable expenses in going to work it is quite 'rational' for both these groups to be voluntarily unemployed. The authors make a comparison with data from 1978: then, they suggest, 8 per cent were certainly better off on the dole and 20 per cent probably so. For the average short-term unemployed the insurance benefits amounted to 58 per cent of previous income in 1981, for the long-term unemployed to 43 per cent.

Averages disguise, of course, as much as they reveal. Income during unemployment depends on several factors: length of unemployment, family status, number of children, additional workers in the household and informal sources of support. One OECD report estimates the average proportion of household income maintained in the second year of unemployment for a married couple with two children as 54 per cent, for a single person as 28 per cent of previous earnings (OECD, 1983).

Another OECD estimate (1982) suggests less dramatic cuts. Unfortunately comparability is handicapped because this estimate takes note only of short-term unemployment; a man whose wages were 66 per cent of those of an average production worker, with a wife and two children, who was unemployed for 6 months in the United Kingdom in 1980–81 maintained, according to this estimate, 93.9 per cent of his previous income during a full year which includes 6 months of paid work.

There are other widely varying estimates quoted in OECD documents of the degree of income maintenance during unemployment that bear on the issue of 'voluntary' unemployment based on a financially 'rational' decision of individuals. The extent of such voluntary unemployment is, according to OECD's cautious evaluation of conflicting data and assumptions, less than often assumed: 'There is little convincing evidence that voluntary unemployment is of any sizeable importance . . . present high levels of unemployment have to be viewed for any practical purposes as mainly involuntary' (1982, p. 9).

This conclusion receives indirect support from experience since the introduction of more stringent criteria for registering as unemployed and from the abolition of earnings-related benefits in the United Kingdom, both measures reducing the amount of benefit for some. If income maintenance were a major factor in remaining voluntarily unemployed, the overall rate of unemployment should have been affected since 1978 when, according to Kay and Morris (1982), 28 per cent were probably financially not much worse off than in employment.

So far as the United Kingdom is concerned this conclusion is strongly supported by the evidence of Micklewright (1985) on the actual benefits received by a sample of unemployed workers:

. . . overall the picture is quite clear: the unemployment benefit system in the 1970s was much less generous than many people have supposed. Changes to the system in the last five years can only have reduced still further the level of support provided by benefits. There are strong reasons for doubting whether the present government's pre-occupation with the disincentive effects of unemployment benefit was ever justified. Now, more than ever, questions should be asked about their adequacy.

Despite some recent simplistic arguments to the contrary, the explanation for the current level of mass unemployment does not lie in voluntary withdrawal from work. There is widespread agreement among most economists that a multiplicity of causes must be taken into account. We go along with Nickell's sarcastic opener in reviewing Minford's 1983 book, *Unemployment: Cause and Cure*:

For one who has been raised in the 'on the one hand, on the other hand' school of economics, it is refreshing to read a book the content of which lives up to the breathtaking certitude of its title. Question: what is the cure for unemployment? Answer: (i) a maximum ratio of 70% for total unemployment benefits relative to net income in work; (ii) the common law to apply to all union actions; no closed shop and a Labour Monopolies Commission. These are the key unemployment-reducing policies which will lead to a long-run fall in unemployment of 1.7 million. A powerful cure indeed. (Nickell, 1984, p. 946)

and his suggestion that such a quick and easy way out solution is based on very little factual evidence.

This more sanguine economics debate today forms part of the so-called 'new classical' macroeconomics and the 'rational expectations hypothesis' with which it is often associated. The major impetus for their increasing influence was the coexistence of high unemployment and rapid inflation in the late sixties and early seventies; the notion of a stable trade-off between these factors (the Phillips curve), which had strongly influenced policy in the sixties no longer appeared realistic. The assertion of Friedman (1968) that there exists a 'natural rate' of unemployment consistent with given real wages, suggested that, in the long run, this trade-off was illusory – an attempt by government to reduce unemployment by monetary or fiscal policy could be successful in the short run, but only at the expense of higher inflation which would then be built into the system. On realizing that an increase in money wages was not matched by an increase in real wages, some workers would cease to offer their services and unemployment would rise to previous levels. Further, workers would subsequently base wage demands on (revised) expectations of inflation. A given level of unemployment is now associated with higher inflation, which continues to accelerate while unemployment is forced below the natural rate. In the long run, the Phillips curve is vertical – there is no trade-off between inflation and unemployment.

The rational expectations school developed these ideas further. In essence, the hypothesis of rational expectations asserts that individuals act as if they knew the structure of the economy and form expectations according to all available information. Government policy is ineffective if known to economic agents: if the policy is anticipated, behaviour is adjusted to neutralize its effects (see, e.g. Begg, 1982 for a full discussion).

In the context of the relationship between inflation and unemployment, this implies that even in the short run a reduction in unemployment cannot be achieved by, for example, systematic monetary policy, since there is no reason why the workforce should not foresee that its effect on real wages would, on average, be neutral. Only lack of information and forecasting errors – which could be in either direction – would cause unemployment to deviate from the natural rate.

It appears to be historical accident rather than inherent theoretical necessity that the idea of rational expectations has been embraced mainly by economists whose thinking is dominated by the notion of a perpetual tendency for markets to clear. In combining that tendency with rational expectations, these economists are forced to conclude that unemployment is essentially voluntary. Robert Lucas, one of the leading proponents of the new classical macroeconomics, is cited in Sheffrin (1984) as follows:

Involuntary unemployment is not a fact or a phenomenon which it is the task of theorists to explain. It is, on the contrary, a theoretical construct which Keynes introduced in the hope that it would be helpful in discovering a correct explanation for a genuine phenomenon: large-scale fluctuations in measured, total unemployment. Is it the task of modern theoretical economics to 'explain' the theoretical constructs of our predecessors?

One wonders whether someone who regards involuntary unemployment as a 'theoretical construct' lives in the real world. To assert that measured unemployment and involuntary unemployment are totally separable, flies in the face of all the evidence. At a minimum, one would need to show that a 'voluntary holiday' was more attractive in 1985 than in, say, 1975 or 1965. As discussed before, there is nothing to suggest that unemployment is now more economically attractive than at other times over the last one or two decades.

In an interview with R. M. Solow, Klamer (1984) records the former as follows:

Suppose you ask a new classical economist, 'what is the central problem of macro-economic theory?' I think the answer would be 'to explain how nominal disturbances can have real effects'. New classical economists wonder why rational people do not always see that a 10 per cent increase in the money supply is equal to a 10 per cent increase in every price overnight and that all real things remain in the pre-existing equilibrium. . . . It seems to me that the formulation of nominal shocks, that Lucas and Barro prefer, is favourable to their own view of the world. . . . But there is no way that you can say that if a real shock occurs, say the invention of television or the computer, everyone in the economy could be able to figure out what the new equilibrium is. . . . No one in his right mind could believe that everyone can figure out how such inventions change real wages, relative prices and all that.

As will become apparent in the following chapters, it is central to our analysis that expectations are important. But unlike most of those in the new classical tradition, and in sympathy with Solow's remarks, it is expectations generated by real shocks to the economy, and in particular the opportunities offered by innovations, that are of significance to the problem of unemployment. The question of a trade-off between unemployment and inflation has simply not been put to the test in the current environment. Analogies with the thirties would suggest that moderate reflation carried minor inflationary risks; analogies with the seventies suggest that it would carry greater risks, but it is unclear whether the experience of the more recent period is more appropriate to the immediate situation, with unemployment close to the prewar levels.

However, this is not the place to pursue this economic policy debate. We will return to it in part IV and the conclusions to this book. Our task here is to relate this brief economic analysis of the unemployed to an examination of the psychological and social consequences of unemployment (for a fuller discussion see Jahoda, 1982).

The Experience of Unemployment

For the last 200 years or so employment has been the dominant social institution through which people earn their livelihood. So powerful is the institution of employment that it has significantly shaped the way of life in industrialized societies. It has led to the establishment of other institutions, such as the trade unions and employers' organizations; it has shaped the routines of family life of the employed, employers and self-employed, the organization of public transport and leisure facilities and other features of modern life.

Like all social institutions employment has manifest purposes and latent consequences. The manifest purposes vary with the point of view adopted. From the perspective of the collectivity the purpose is the production of goods and services beyond those that independent individuals can provide for themselves; from that of the employer it is to make a profit; from that of the employee to earn a living. The pursuit of these interdependent purposes has given the institution of employment its characteristic shape.

Beyond these manifest purposes employment has latent consequences. It is so organized that it makes certain types or categories of experience inevitable (Jahoda, 1982). Whether people love or hate their jobs, employment shapes their experience of the passage of time for the day, the week, the year and a lifetime; employment inevitably enlarges the social horizon through contacts beyond the family and self-chosen friends; in employment people experience themselves as part of a collective effort, contributing to and depending on the work of others; through their jobs they are assigned a social identity that defines their place in society; and finally they are compelled to engage in some activity whose purpose goes beyond the pain or pleasure they experience while performing it. The quality of the experiences within these categories is still too often negative, even though conditions of employment are now more tolerable than they were in the brutal phase of early capitalism. To improve the quality of experience within these categories should be on the agenda of all efforts to raise the quality of working life, but this is one of several major tasks pushed into the background by the even greater problem of unemployment.

Systematic studies of the experience of unemployment (Miles, 1983) reveal that the psychological debilitation of a large majority of the unemployed is the result of deprivation in the categories of experience that employment provides: unstructured time becomes a heavy burden rather than leisure which is a complement to employment, not a substitute for it; many unemployed feel isolated and cut-off; they resent their enforced uselessness and exclusion from participation in collective purposes, feel abandoned by society, without social identity, on the scrapheap; and inactive and bored when there is nothing that needs to be done.

Employment is, of course, not the only institution that makes such experiences possible. For better or worse, however, it has come to be the major one that meets these human needs on a regular basis. The manner in which employment structures these experiences is often and under-standably resented. Only when the loss of a job removes the automatic, taken-for-granted provision of experiences do people become fully aware of their psychologically supportive function.

To say that work in employment meets some basic human needs is not to suggest that it meets all. People need other things too, among them just the opposite of what employment provides: unstructured time, privacy, individual efforts, self-esteem independent from social norms, and periods of play or inactivity. These two sets of needs are complementary. It is an empirical fact, not a moral judgement, that most people in the modern world are better able to satisfy their leisure needs if those provided by work are met first. This one concludes from the frequent suffering that accompanies retirement even though it conveys a respected social identity; or from the use of tranquillizers or alcohol among many housewives when their children have left home, to help them over the absence of useful work even when there is no financial strain; but above all from the psychological state of the large majority of the unemployed. Among all three groups – the retired, underoccupied housewives and the unemployed – there are some who do not suffer so, because they have out of their own initiative undertaken some voluntary work that gives them much of the psychological support that employment used to provide. To expect that the majority of the unemployed could do likewise amounts to psychological utopianism. In the European community 40 per cent of the unemployed are under 25 years of age; a large majority of the adults are unskilled or semi-skilled and come from an educationally and financially underprivileged background – not conditions conducive to the development of the initiative and idealism required for regular engagement in voluntary work.

Two overlapping groups among the unemployed – the young and the rising number of long-term unemployed – give special reason for concern. Some of the young still live more or less resentfully with their parents and, provided that at least one of them has a job, do therefore not lack subsistence. All of them, however, are deprived of much that goes with being young: striving for independence, hoping and planning for the future. Even before leaving school the prospect of unemployment dampens their aspirations and their motivation to learn. Once they are on the dole all too many abandon their desires for something better. If society treats them as outcasts, they will live as outcasts from day to day without a thought for tomorrow. Such adaptation to being unemployed as a life style may reduce their psychological suffering for the moment. They and society will have to pay for it later.

As long-term unemployment increases, so does the plight of individuals. The financial pressure intensifies; a year or more of unemployment operates against re-employment, particularly for those aged 45 or above. Even the lucky ones who find another job often fall victim to the principle of 'last in – first out'. Repeated spells of unemployment increase the number of discouraged workers.

The young and the discouraged unemployed swell the number of people who form what the Americans call an 'underclass': permanently dependent, without hope, demoralized.

The deficit in psychological well-being among the unemployed is, however, not limited to these two groups (Warr, 1983). The degree of disturbance and the manner in which it expresses itself varies, of course, from individual to individual with personality, life history and many other factors. The dominant response is surprisingly similar to that documented for the thirties: depression, frustration and resentment against enforced uselessness. The similarity of experiences during unemployment over a period of 50 years during which enormous social changes occurred supports the notion of the psychological significance of employment in providing the traditional environment in which basic human needs are met.

Social Consequences

When frustration, depression and resentment of an individual is multiplied by several millions there must ensue consequences for the entire society over and above the economic damage inherent in mass unemployment. It is, however, immensely difficult to provide systematic evidence for the ethos of complex societies, such as Britain, and even more difficult to establish the impact of unemployment on it. The most comprehensive effort in this direction was made by Harvey Brenner (1976) who correlated various indicators of social pathology – morbidity, mortality, crime, suicides, etc. – with unemployment rates in the United States, parts of Britain and Sweden. In all three countries he found these indicators deteriorating after various time-lags with an increase in the rate of unemployment. His impressive effort has been criticized on methodological grounds. Notwithstanding the controversy about his work, which has an ideological undertone, he may well be right, though estimates of social consequences of unemployment have probably an even wider margin of error than estimates of economic factors.

Street violence, crime and the black economy have all been blamed on unemployment, but there is only sparse and contradictory evidence for such claims. To the extent that such activities are organized on a semi-permanent basis, they do indeed meet some of the needs frustrated by unemployment. Their organization requires, however, a degree of initiative and, in the case of the black economy, the possession of skills and tools that are largely absent in the life situation of the unemployed and contrast sharply with the passively despondent mood among most of them. Acts of vandalism or occasional days of unreported paid casual labour by unemployed people are more likely. But there is no reason to believe that they are undertaken on a scale likely to disrupt or disturb the entire society.

Less visible to the naked eye than the symptoms of social pathology captured in indices, but equally threatening, is another social consequence of mass unemployment: a polarization in the population not according to social class but according to employment status. If that continues for many years it may undermine that minimum of consensus that is required for the maintenance of a civilized society.

Public opinion polls report that the vast majority of the population regard unemployment as the major social problem of the decade. There are, however, indications that this general concern with unemployment should not necessarily be interpreted as an expression of solidarity with the unemployed nor of compassion. Several studies have investigated what employed people think about the unemployed (e.g. Furnham, 1982; or Breakwell et al., 1984). Those still in employment often tend to blame the unemployed themselves for their personal fate. Among the unemployed there is less self-blame; they hold the general economic situation or government policies more often responsible for job losses.

This trend toward a polarization in attitudes is paralleled by a significant polarization in living standards. The income gap between the two groups is growing, since on the average, wages and salaries are keeping ahead of inflation while unemployment benefits are falling behind. Together with the defensive attitude of either group to the other this may produce rejection and contempt for the unemployed among the employed, and envy and resentment among the unemployed. There can be no doubt that unemployment is one of the main factors also in the rapidly widening gap between rich and poor in the United Kingdom (*Social Trends*, 1986).

Fear of unemployment polarizes also attitudes toward the introduction of new technologies in the entire society. While many industrialists, trade union leaders and people within government regard it as necessary for economic recovery, there is in the labour force an understandably widespread reluctance to come to terms with the new technologies which they see mainly as labour-saving, not as job-creating. The introduction of radical change is never a painless process. If accompanied by the fear of unemployment it leads to strong resistance and disputes which interfere with the potential of new technologies for job creation as well as productivity increases. The spectre of unemployment induces a concentration on the narrow self-interest of relatively small groups of workers that leads to intra- and inter-union conflict, and often to conflicts between shopfloor, shop stewards and trade union leadership.

One example of such conflict concerns overtime. It persists even though some central trade union leaders advocate its curtailment as one step among others in the effort to reduce unemployment. The attraction of overtime for low paid workers and the lack of solidarity with the unemployed have so far prevented progress in that direction.

In the long and difficult run a radical shortening of working time is one possible solution to unemployment. A working day of 6 hours would provide the psychological support that employment offers just as well as an 8-hour day and might ease the way for all who want to work to find jobs. Furthermore, these jobs would be less exhausting and their unsatisfying features would be more tolerable during a shorter day. Compared with the beginning of the nineteenth century working time has been cut by 50 per cent. A further 25 per cent reduction should not require as long, notwithstanding the economic difficulties such a radical change would imply. However, these transitional difficulties are sufficiently great in a competitive international economy that this road cannot offer any instant solution.

Other suggestions for shortening working time propose earlier retirement or late entry into the labour force. Retirement at 55 or 60 years of age would indeed have advantages for the young in terms of job availability, but if employment fills deep-seated needs it would deprive the older age group by the same token, perhaps with the exception of some professional workers. In addition, early retirement could do nothing to improve the quality of working life of those with jobs.

Later entry into the labour force has much to commend itself in theory; in practice it would presuppose a radical change in current educational institutions, where teachers often find it difficult to keep some children interested even to the current school-leaving age. Nevertheless for others it could bring great benefits. Educational periods interspersed with periods of work would probably often be more effective and more appreciated. Certainly policies for education, training and retraining are a crucial part of the solution to our problems and these are taken up in chapter 12. In this case there is a happy coincidence between social and educational goals and the need to acquire skills of new technologies.

The way to a socially more coherent and psychologically more satisfying society will be found easier if policies are based on an understanding in depth of the economics of employment and unemployment. The research results presented in the following chapters were prompted by the conviction of the potentially destructive impact of unemployment, both on individuals and on the entire society. They are presented in the hope of contributing to ways of reducing and ultimately perhaps eliminating this scourge on modern life.

2
Employment, Unemployment and Technical Change: A Review of the Economic Debate

Luc Soete

It is not the intention here to review fully the literature on the relationship between technical change, employment and unemployment. Thorough reviews of this literature can be found in Gourvitch (1940) and more recently Cooper and Clark (1982), Fano (1984) and Katsoulacos (1985). In order to set out our own approach better, it might be useful to summarize briefly some of the leading contributions of the main schools of economic thought. We shall do this under four main sub-headings: classical, neo-classical, Keynesian and 'structuralist', concluding with a brief excursion into the contemporaneous debate.

The relationship between technical change and employment was at the forefront of economic policy debate long before 'economics' emerged as a 'science' with the appearance of Adam Smith's *Inquiry into the Nature and Causes of the Wealth of Nations* and the emergence of the 'classical' economic school of thought. The origins of the debate can be traced back as early as the sixteenth cntury when 'awareness' of progress and its accompanying employment displacement started to influence mercantilist policy and gave rise to a number of mercantilist writings. These were often contradictory in nature and can be considered as indicative of the continuous unease with which many economists would later on approach the debate. Thus whereas 'the facilitations of Arts' were generally viewed as greatly beneficial to society (Petty, 1899, p. 118) concern over the employment displacement effects was great and often led to legislation restricting the use of machinery (Heckscher, 1935, p. 172). With the further emergence of the industrial revolution and the dramatic dislocation effects on peasants, craftsmen and workmen of all sorts – often resulting in the violent destruction of new machinery (Clark, 1984) – the employment implications of the introduction of new machinery became a crucial point of debate in most economic writing.

The Classical School

All schools of economic thought have recognized that the issue of labour-displacing technical change must be seen in the context of the economic system as a whole. They have stressed the importance of 'compensation mechanisms' in the system which may generate new employment to 'compensate' for the loss of old jobs, but they differ greatly in their assessment of the speed, efficacy and determinants of these mechanisms.

In general it could be said that the classical economists brought clarity in the debate by pointing to the significant employment 'compensation' mechanisms resulting from the introduction of new technology, which at least in the long run would be operating in the economy at large. Even before Adam Smith's *Wealth of Nations*, James Steuart had given a clear and concise interpretation of how 'sudden' mechanization could lead to temporary unemployment, compensated in the long run by employment growth in the machinery producing sectors and overall compensating effects due to price reductions. Whereas Steuart recognized that a demand for goods did not in itself imply a demand for labour, it was Ricardo who, after acknowledging the 'general good' technological improvements would bring about, 'accompanied only with that portion of inconvenience which in most cases attends the removal of capital and labour from one employment to another' (Ricardo, 1821), shocked his readers and provoked a 'furore of debate' by granting that:

The opinion entertained by the labouring class, that the employment of machinery is frequently detrimental to their interests, is not founded on prejudice and error, but is conformable to the correct principles of political economy.

Although he later qualified his view and much debate still surrounds the interpretation of his ideas, Ricardo's suggestion that the introduction of labour-saving machinery might be injurious to the workers, causing unemployment when real wages could not be forced downwards or local machine production expanded sufficiently, was fundamentally based on his recognition of lags and inflexibilities in the employment compensation mechanism, resulting directly from supply-side constraints. These rigidities and time-lags remain a central issue in the present debate on structural adaptation to technical change.

By focusing on the supply of capital as the crucial factor in bringing about employment 'compensation', the classical economists implicitly assumed a system of fixed coefficients. Whereas neo-classical theory has tended to stress the ease of substitution between labour and capital (i.e. to look upon capital as 'putty' rather than 'clay'), many economists continued to view the rigidity of the capital stock, especially in manufacturing industry, in much the same way as the classical economists. Thus, for example, Robertson (1931) pointed out:

. . . the tendency to industrial rationalisation is, in one of its aspects, a tendency to install such elaborate and expensive and durable plant, and to devise such a close and intimate co-ordination between it and the labour force required to work it, as to leave as little room as possible for the operation of the Principle of Substitution.

Similar arguments to those of the classical school with regard to the importance of 'embodied technological progress', fixed as 'clay' – in new investments – can also be found in some of the present-day 'vintage' growth models. Salter (1961) in particular, very much in Ricardian tradition, went as far as comparing capital with land:

Capital goods in existence are equally as much a part of the economic environment as land or other natural resources. Both are gifts: natural resources are the gift of nature, capital goods are gifts of the past. The fertility of land corresponds to the level of technical knowledge embodied in capital equipment . . . in both cases the margin of use is determined by the ability to show a surplus or rent; no-rent land corresponds to capital equipment that has just been abandoned.

What is interesting about these more recent vintage growth models (Driehuis, 1979; Arthus, 1983) is that here too, there is a clear possibility of capital shortage unemployment because of labour-saving technological change, as in the classical economists' underlying trend towards 'mechanization' or the conversion of 'circulating' into 'fixed' capital (Ricardo, John Stuart Mill) or of 'variable' into 'constant' capital (Marx).

The importance given to capital supply in classical economics, as well as the fixed coefficients assumption, reflected an 'awareness' of the complexity of technical change, and particularly of its 'sudden', even revolutionary nature.

For classical economists, from Smith to Ricardo, John Stuart Mill and most of all Marx, technical progress was indeed one of the crucial variables in the economic system. Marx in particular emphasized the social transformation brought about by 'technological revolutions'. Thus gunpowder had put an end to knightdom; the compass had opened world markets and printing had brought about the renaissance of science. Capitalism was characterized by a restless search for new products and processes of production, indeed – 'the bourgeoisie cannot exist without constantly revolutionizing the means of production'. In his model of the long-term cyclical pattern of the growth of the system, the constant process of embodying new technologies in new types of capital equipment led to a growth in capital intensity (the 'organic composition of capital') to a fall in the rate of profit and to a tendency for unemployment (the 'industrial reserve army') to grow larger with each successive cyclical crisis.

There is little doubt, that as much in their awareness of the 'suddenness' of technological progress, as in their recognition of the possibility of – at least temporarily – 'technological' unemployment, most classical economic writing simply reflected the acute conditions of the time: the dramatic technological transformations and the distressful labour conditions particularly in the period from 1815 to 1840. With the rapid growth of most economies during the second half of the nineteenth century and the 'Belle Epoque' before the First World War, technological progress came to be regarded as 'gradual and continuous' rather than 'revolutionary'. This also was more congenial to the marginalist assumptions of the neo-classical school in their theories of production and consumption. As a consequence, economists' perception of the problem of technological progress and employment displacement changed quite fundamentally. As Gourvitch (1940, p. 30) points out: 'As far as economic literature was concerned, the problem of technological unemployment virtually ceased to exist; it did not reappear until the 1920s.'

The Neo-Classical School

In classical economics technological unemployment could arise because the market could fail to absorb fully the increased production resulting from technical change and because capital supply could be insufficient to absorb the displaced labour. In neo-classical economics, the market price mechanism will take care of both. General overproduction, particularly as the result of technical change, will in principle be ruled out by the self-regulating commodity price mechanism. To quote Marshall (1920): 'The net aggregate of all the commodities produced is itself the true source from which flow the demand prices for all these commodities, and therefore for the agents of production used in making them.'

Say's law was thus accepted and indeed became the cornerstone of neo-classical economics from Jevons, Marshall, the Austrian school and onwards. Whereas overproduction of particular commodities may occur, this will be accompanied by underproduction in other commodities. As Gourvitch puts it: 'There can be no general overproduction of commodities, no general oversupply of all commodities in relation to a given system of prices. . . .'

In relation to capital supply, the possibility of the pace of technological change 'outrunning' capital accumulation is ruled out through the possibility of factor substitution between capital and labour mediated by the factor price mechanism; the wage and interest rate assuring that no factor of production can remain 'unemployed' in the long run. Factor substitution is the main feature distinguishing neo-classical from classical economic thinking. According to Marshall (1920) 'The principle of substitution' will bring about the best possible combination of capital and labour, i.e. that in which both labour and capital obtain the greatest marginal return:

As far as the knowledge and business enterprise of the producers reach, they in each case choose those factors of production which are best for their purpose; the sum of the supply prices of those factors which are used is, as a rule, less than the sum of the supply prices of any other set of factors which could be substituted for them: and whenever it appears to the producers that this is not the case, they will, as a rule, set to work to substitute the less expensive method.

Furthermore, if there is excess of any one production factor, it can always be used to produce something if the correct set of prices prevails, and as it is assumed that overproduction cannot occur, unemployment cannot occur. In other words, just as a balance of supply and demand will be secured in the commodity market through the price mechanism, so will the prices of labour and capital secure the balance of supply and demand for labour and capital. In other words, unemployment cannot occur in the long run, because there will always be a wage rate which will clear the labour market. In von Mises' (1936) words:

. . . lack of wages would be a better term than lack of employment, for what the unemployed person misses is not work but the remuneration of work. The point is not that the 'unemployed' cannot find work, but that they are not willing to work at the wages they can get in the labour market for the particular work they are able and willing to perform.

On the capital side, capital accumulation will bring about a decline in the interest rate, and substitution of capital for labour. As additional quantities of capital are substituted for labour, they will, however, be substituted in a less-and-less profitable way. In other words, the marginal productivity of capital will be declining. At the same time the fall in the interest rate which initially had stimulated investment demand will reduce savings, i.e. the supply of capital will be reduced, until the price of capital, the interest rate, restores the balance between the demand and supply of capital.

This neo-classical general equilibrium framework can be said to correspond most closely to present-day traditional economic views on technical change and employment, Technological change may indeed result in some temporary unemployment, but with efficiently operating commodity and factor markets, there is no basic economic problem arising from the introduction of new technology. Two further points should be added to this:

First, that technological change is seldom viewed as 'sudden' or revolutionary, but rather that the introduction of new technology and its diffusion is seen as a slow and gradual process, taking decades rather than years. Even confronted with major commodity and factor price rigidities, the ensuing unemployment will not have anything to do with technological change.

Second, that even in cases of so-called 'non-neutral', 'biased', labour-saving or capital-saving technical change, this will generally speaking be 'induced' by changes in relative prices, stimulating both the research and discovery of new production methods and the application of technologies which might have been profitable even before the relative price change, but were not used (Hicks, 1935, p. 125).

Both points lend further support to the argument that it is the failure of wages to adjust to the labour market clearing price or alternatively to the decline in the marginal productivity of labour in the case of labour-saving technical change, which is the main cause of structural unemployment.

The Keynesian School

The Keynesian school can probably best be distinguished from the neo-classical school by its rejection of the notion that equilibrium necessarily implies full employment. Whereas Say's law does indeed hold in the case of full employment, it does not in any equilibrium position with 'underemployment'. These are truly equilibrium positions in the sense that they do not automatically lead to any movements which would bring about full employment. Such positions are characterized by insufficiency of effective demand. The latter itself is intrinsically linked to the two concepts which distinguished Keynes most clearly from previous economic orthodoxy: the marginal 'propensity to consume', which is assumed to decline with increases in aggregate real income and the 'liquidity preference' limiting the expansion of investment. It is not the place here to go into a detailed description of the main contributions of Keynesian thinking in the area of macroeconomics. Suffice it to say that Keynes denied both interest and wages the self-regulating equilibrium functions assumed in neo-classical economics.

Thus rather than increasing the demand for labour, a reduction of money-wages in the case of 'demand-deficient', cyclical – today often referred to as 'Keynesian' – unemployment will reduce further effective demand, exacerbating the unemployment problem rather than solving it. Likewise, interest rates in so far as expressing as much a desire to hold wealth in the form of cash – the liquidity preference – as the reward for 'waiting', will not reflect the equilibrium price of the demand for new investments and the supply of savings.

What then determines investments? Keynes was notoriously sceptical about the rationality of investment decision-making and in a well known and often quoted passage in *The General Theory of Employment, Interest and Money* (1936) compared investment with an expedition to the South Pole, since the uncertainties affecting the future stream of anticipated profits were so great. Technological progress could undoubtedly be considered as one of those major uncertainties even if Keynes remained at least, in *The General Theory*, surprisingly silent on the subject. It is not without interest, however, that some years earlier (in 1930), Keynes did acknowledge the great importance of Schumpeter's theory of waves of investment stimulated by major technical innovations:

In the case of fixed capital it is easy to understand why fluctuations should occur in the rate of investment. Entrepreneurs are induced to embark on the production of fixed capital or are deterred from doing so by their expectations of the profits to be made. Apart from the many minor reasons why these should fluctuate in a changing world, Professor Schumpeter's explanation of the major movements may be unreservedly accepted.

More generally it can be said that Keynes did not carry further his inquiry into the relationship between technological change and employment. Whereas indeed he related unemployment directly to underinvestment – through liquidity preference – he did not follow the classical economists in viewing investment growth at the expense of consumption; capital shortage was a short-term, monetary phenomenon, it was not related to long-term 'mechanization'.

His rare statements on the long term, particularly in relation to the long-term tendency of the marginal efficiency of capital to fall as accumulation would proceed, did however fit classical economic thinking (the tendency of the falling rate of profit). But it is worth noting that in contrast to Marx, Keynes viewed the long-term decline in capital productivity as the achievement of capitalism's ultimate goal. It meant 'the euthanasia of the cumulative oppressive power of the capitalist to exploit the scarcity value of capital' and the exchange of commodities 'at a price proportional to the labour etc. embodied in them'; it reflected the satisfaction of all consumer 'wants', rather than capitalism's ultimate crisis (Keynes, 1936).

Elaborations on Keynesian thinking in the long run can be found in the early growth literature and particularly in the contribution of Harrod (1948). Harrod extended Keynes' ideas to the case of a growing economy. His conclusion, that the economy is inherently unstable, and that even in case the 'warranted' stable growth rate would be achieved, there is no reason to expect it to coincide with the 'natural', full employment growth rate, is reminiscent of classical economics. It brought about an explosion of research and debate on the growth dynamics of the system – the emergence of modern growth theory, and the capital debate – which, at least from our perspective, seems however to have focused almost exclusively on the identification of so-called long-run 'steady state' full employment growth paths.

Thus Solow's (1957) most influential neo-classical growth model, by introducing factor substitution and equating the level of investment with the level of savings associated with full employment, did indeed show that a steady state (full employment) growth path would always exist in the economy, in the sense that if initially the conditions for a stable growth path do not hold, the economy will always adjust towards it. However, it could be argued that in this formulation the question of how technological change affects the stability of economic growth has been obscured rather than clarified. The representation of 'smooth shifts' of the whole production function does not conform to the engineer's or technologist's perception of technological change, as we shall see in chapter 3. It can therefore be said that growth economists by and large side-stepped the issue of the impact of technological change on full employment.

In their review of the literature on technical change and employment Cooper and Clark (1982) observed, in assessing the contribution of neo-classical growth models to the debate:

It is probably fair to say that such approaches to technological change partly account for the limited amount that macro-economics has had to say on many practical questions which policy-makers and others feel to be important. It is true, of course, that economists concerned with policy matters within a fairly short time horizon have been concerned with problems of disequilibrium – particularly unemployment – which technological change might cause; it is also true that as old 'verities' are increasingly put in question, particularly regarding government's ability in demand-management, some of the optimism about the instability problems which technological change might cause has been attenuated. But it is still possible to argue that the emphasis in the growth literature on long-run steady states of growth and full employment dominates some economists' thinking on technological change and restricts quite severely the questions they are inclined to ask about it.

The Structuralist School

Under this heading we will focus primarily on Schumpeter, but include also some comments on Lederer, who made some particularly interesting observations on changing industrial structure and employment.

In contrast to Keynes and neo-classical economic thinking, technological progress was for Schumpeter, probably more than any other economist, at the centre of the dynamics of the economic system. Whereas in the previous two schools of thought, growth was simply accompanied by the emergence of new industries and technologies, for Schumpeter the system was driven by such technical innovations and their diffusion.

Schumpeter justified on three grounds his view that the process of innovation was a major source of disequilibrium in the economic system, rather than a smooth and incessant type of transformation. First, he argued that innovations are not at any time distributed randomly over the whole economy, but tend to concentrate in certain key sectors and their immediate environment, so that they are by their very nature lop-sided and dis-harmonious and very often give rise to problems of structural adjustment between different sectors of any growing economy. Secondly, he argued that the process of diffusion, through which innovations bring about major surges of investment and of output growth, was itself inherently an uneven process with cyclical characteristics. Whereas the introduction of any new product is often characterized by a slow and hesitant first start or a series of false starts whilst teething troubles are overcome, this is usually followed by a rapid growth phase. In Schumpeter's (1939) analysis this phase of very rapid growth was associated with the 'swarming' or 'band-wagon' effects of imitation, as a crowd of firms attempted to exploit the new opportunities for profitable investments and market growth, which are now widely perceived.

Finally, Schumpeter stressed that profit expectations would change during the period of rapid growth, because the swarming process would tend to erode the profit margins of the innovators. As new capacity was expanded, at some point growth would begin to slow down. Market saturation and the tendency for technical advance to approach limits, as well as competition and pressure on costs of inputs, would all tend to reduce profitability and with it the attractions of further investment. Sometimes this whole growth cycle might take only a few years, but for some very important new products and technologies it might take several decades.

Schumpeter maintained that these characteristics of innovations and their diffusion were sufficient to bring about major disturbances – 'gales of creative destruction' – in the economy.

With regard to employment, Schumpeter (1939) ascribed to these major technological transformations the widespread emergence of 'cyclical' technological unemployment:

[Economists] have a habit of distinguishing between, and contrasting, cyclical and technological unemployment. But it follows from our model that, basically, cyclical unemployment is technological unemployment. . . . Technological unemployment . . . is of the essence of our process and, linking up as it does with innovation, is cyclical by nature. We have seen, in fact, in our historical survey, that periods of prolonged supernormal unemployment coincide with the periods in which the results of inventions are spreading over the system and in which reaction to them by the system is dominating the business situation, as for instance, in the twenties and in the eighties of the nineteenth Century.

Lederer's (1938) analysis of the relationship between technological change and employment appears in many ways the logical extension of Schumpeter's analysis. In Lederer's view (labour-saving) technological progress will act as a drain upon capital, not because of any 'classical' conversion of circulating or variable into fixed or constant capital – see our discussion of the classical school above – but because it will involve a diversion of capital from the 'static' to the 'innovative' enterprises. In so far as capital investment can only come from savings, there will be scarcity of capital which would otherwise have brought about the steady and proportional expansion of all enterprises. In the dynamic enterprises output and employment growth will slow down, and combined with the absolute decline in the static enterprises (only labour-saving technological change is considered), this might lead to increased structural unemployment. The main reason why the employment displacement arising from technological progress will be 'disguised' and deferred till later stages of the ensuing economic boom has to do, as in Schumpeter, with the possibility of financing the introduction of the new technology through 'credit creation'. As in Schumpeter the result is a clearly cyclical relationship between technological change and employment growth. In the words of Gourvitch (1940, p. 172):

The limitations placed upon employment by the lack of capital are made inoperative for a time. . . . Expansion of technically stationary enterprises, together with growing investments in the technically progressive units, will go on [through further cumulative credit expansion] until inevitable deflation sets in. The displacement of workers will thus be 'postponed' until the depression stage. 'Cyclical' unemployment will then be combined with 'structural' unemployment as a consequence of technological progress.

The Contemporary Economics Debate

It is perhaps an overstatement to talk about a 'contemporary economics debate' in the area of technical change and employment. With the exception of the US automation debate in the sixties there has been a gradual, near complete, disinterest in the subject of technological unemployment on the side of the academic economics community. The number of academic articles, such as Neary (1980), Sinclair (1981), Stoneman (1983), Katsoulacos (1984, 1985) and Venables (1985), over the last decade is very small, by comparison with the thirties when the issue was at the forefront of both academic (Douglas, 1930; Hansen, 1932a/b; Gregory, 1931; Kaldor, 1932; Bourniatin, 1933; Neisser, 1942; Weintraub, 1937) and economic policy (the NBER studies such as Jerome, 1934; Mills, 1932; Gourvitch, 1940; Fano, 1984) debate in the US and Europe alike.

Part of this neglect is in our view simply a reflection of the more widespread general neglect of the analysis of the technical change factor in macroeconomics over the last decades and the focus on general equilibrium as the main approach to the understanding of the operation of economic markets and economic agents. From a general equilibrium point of view the technology employment issue reduces to a simple distributional question. In so far as technological change increases factor productivity this will be reflected in an increase in real aggregate income. The only relevant question with regard to employment within such a framework amounts to how this increase in real income will be distributed and spent. In cases of labour market failure the increased real income could indeed become wasted in the form of unemployment for the 'marginal' workers.

A major dynamic factor of disequilibrium such as technological change, not just corresponding to 'factor augmentation', but involving the creation of new products and the rise of entirely new industries; the process described by Schumpeter as one of 'creative capital destruction', does not fit well such contemporary economic analyses. Furthermore its measurement and even its definition is messy (or alternatively has not benefited yet from general economic approval, such as GNP or total factor productivity). There is consequently a broad consensus among most contemporary economists to treat the technology factor as a black box, not to be opened except by scientists and engineers or occasionally by economic historians. So that for the urgent, short-term problems which most economic analysis is directed towards, it becomes indeed 'reasonable' to treat this long-term 'messy' technology variable as essentially continuous and exogenous.

This undoubtedly lies also behind the failure of 'labour economics' to pay much attention to the technology factor. For example, in their recent study of the causes of British unemployment, Layard and Nickell (1985) assume that the actual rate of technical change has been declining over the recent period and consequently can be of little importance in explaining the growth in British unemployment. It is probably fair to say that this statement reflects more the almost automatic dismissal of the technology factor in the contemporary labour economics debate than the authors' detailed analysis of the rate and direction of technical change.

A slightly 'keener' interest in the technology–employment relationship emerged from that branch of economics which focused on technological change, industrial innovation and diffusion and which often found its inspiration in dissatisfaction with the traditional way economics 'assumed away' some of the main characteristics of the economic growth process, especially research, invention, innovation and diffusion. Whereas some of these contributions led indeed to alternative views of the workings of the economic system: such as Nelson and Winter's evolutionary model of technical change (1982); others involved more traditional economic theory elaborations focusing, e.g. on the nature and direction of technical change and its relation with factor markets; such as Kennedy (1964) and von Weizsacker's (1966) invention possibility frontier. Again, it is also probably fair to say that, at least until recently, the employment–technical change relationship did not emerge as an issue of prime interest, even in the economics of technical change.

Two more or less distinct types of recent contribution can nevertheless be distinguished: first of all, those which fit more or less within traditional neo-classical economic theory and have tried to specify in a more formal way the conditions under which full employment compensation would occur, and second those which, building upon some of the Schumpeterian insights in the process of innovation and diffusion, have tried to relate these disequilibrium features of the process of technological change in a more systematic way to long-term trends in employment growth.

As regards the first type, despite what the reader might be tempted to believe following our historical survey, the conditions under which 'employment compensation' will occur within a traditional neo-classical setting are far from easy to specify, once the existing interindustry flows are taken into account, and involve a set of complicated interactions, depending on the industry values of substitution and demand elasticities, rates of factor augmentation with regard to the new technology, movements in factor prices, diffusion speed, etc. To illustrate the high degree of complexity involved, we report here briefly the main 'intuitive' results obtained in a simple, but illuminating model developed by Stoneman (1984) (two factor inputs: capital and labour; factor augmenting technical change; constant returns to scale; constant elasticity of demand and a fixed number of identical firms introducing a new technology at the same time):

(a) If the industry elasticity of demand is greater than unity, then technological change of any type is in the majority of cases going to increase the demand for *both* inputs [capital and labour].

(b) If the industry elasticity of demand is less than unity, then technological change of any type is in the majority of cases going to decrease the demand for both inputs.

(c) Neutral technological change (or biased technological advance with substitution elasticity equal to one) will increase the demand for both inputs if the demand elasticity is bigger than one, decrease it if the demand elasticity is less than one.

(d) There is no presumption that labour-saving technological change will reduce the demand for labour or that capital-saving technological change will reduce the demand for capital except that such effects are more likely if the demand elasticity is less than one.

(e) The effects of technological change depend not only on the absolute sizes of the elasticity of substitution and the elasticity of demand but also on their sizes relative to each other.

(f) A technological advance that is defined as purely saving one factor, affects not only the demand for that factor but also the demand for other factors.

It is worthwhile emphasizing that while these results seem to point to the crucial importance of demand elasticities above one for employment compensation to occur, the most important employment compensation factor: product innovations and the crucial demand growth potential resulting from the associated high income elasticities, is ignored in the model. As Katsoulacos (1984) has shown, whereas new products drive out older ones, new products are more likely to raise the employment potential of an economy both through their direct and indirect demand generating effects. However, his analysis remains primarily a static analysis and does not include the crucial dimension of diffusion.

The dual macroeconomic impact of technological change: the input-saving and demand-creating effects, forms also the focus of the second set of 'contemporary' contributions. What probably differentiates these more technology-focused, Schumpeterian-like contributions most is their rejection of the smooth, neo-classical production function as the main conceptual framework for the analysis of technical change.

As many contributors to the economics of technological change have argued, from Rosenberg (1976) and David (1975) to Nelson and Winter (1982), technological change is typically characterized by strong cumulative effects: 'learning' in its various forms. This means that one will witness a process of 'localized' shifts in the production function to use David's term (1975), of progress along particular 'natural' trajectories, to use Nelson and Winter's concept (1982). The existence of such irreversible dynamic scale economies implies that from an economic theory point of view, non-convexity in the production possibility frontier will appear, with as a result the possibility of multiple equilibria and the occurrence, even the natural tendency towards 'locked-in' technologies (Arthur, 1985) and technological trajectories. We discuss this issue in greater detail in the next chapter, where we set out our own theoretical approach. It is worth noting here though that the theoretical properties of such a dynamic world provide, amongst others, strong support for relatively (short-term) factor price insensitive trends towards particular trajectories or directions of technological progress. 'Labour-saving' technological change, 'mechanization' or 'automation' could well correspond to one such 'natural' trajectory.

The implications of a more rigid, if not (at least in the short term) fixed set of production coefficients for the analysis of technical change and employment are relatively straightforward. From a dynamic point of view labour-saving technological change embodied in new investment could indeed, with slow wage adjustments, lead to unemployment as the result of insufficiency in investment to maintain the full employment capital stock (OECD, 1983), so-called 'capital-shortage' unemployment. Whilst the capital 'deepening' tendency of much investment has been continuing through the rationalization and 'automation' of many processes, the compensating 'broadening' of capital stock through new investment has not been taking place on a sufficient scale. The extent to which in many European countries a large proportion of the increase in unemployment over the last decade could be identified with this 'capital-shortage' unemployment, is investigated in more detail elsewhere (Soete and Freeman, 1985) and with respect to the United Kingdom in chapter 7. We turn first now to the issue of factor substitution and its relevance when dealing with technical change.

3
Factor Substitution and Technical Change

Christopher Freeman and Luc Soete

Following this brief review of economic theory relating to employment and technical change and its relevance for recent debates, we are now in a position to highlight the main differences between our own approach and that of the more traditional, established schools of economic theory. As we have already indicated our approach has more in common with the classical than the neo-classical school, but it derives its more recent inspiration directly from the 'structuralists', who almost alone among twentieth century economists, have tried to look 'inside the black box', as Rosenberg (1982) aptly put it in the title of his recent book.

Of course, no significant body of economic thought, whether classical, neo-classical, Keynesian or structuralist, has ever suggested that the influence of technical change was unimportant. Economists of all persuasions have always accepted that technical innovation was one of the most important, if not the most important source of dynamism in capitalist economies.

However, this superficial appearance of agreement should not be allowed to obscure the fundamental nature of the differences in relation to the analytical framework within which we and other 'structuralists' approach the problems of technology, growth and employment and that adopted by some more orthodox neo-classical and Keynesian economists. We would emphasize the word 'some' because although we have ourselves talked of four 'schools' for purposes of exposition, there are no tidy categories of classification when it comes to assessing the role of technology in relation to growth, investment and employment. For example, Stoneman (1983) although presenting a thorough and sophisticated account of neo-classical theories indicates a personal preference for the theory of Nelson and Winter (1982, p. 195) and many deep reservations about the neo-classical formulation of the problems (see, for example, pp. 54–5).

36

Kennedy and Thirlwall (1973) in their survey of the economics literature on technical change drew attention to the paradox that though all studies of economic growth pointed to technical change as the 'single most important determinant of growth of living standards', there were actually very few studies of the process of technical change itself. They commented that 'the reader of this survey may well have been struck by the apparent thinness of studies in this field as compared to macroeconomic production function studies'. This bias towards labour–capital production functions in the emphasis of mainstream economics has in our view persisted since their survey in 1973, even though the 'aggregate production function' debate has revealed numerous major problems with the concepts, the techniques, the measurements and the assumptions (see, for example Olsson, 1982). Chapter 2 has already pointed to the failure of neo-classical (Keynesian) growth economics to follow up the fundamental questions raised by Harrod (1948). In their review of the economics literature at the start of our own research programme, Cooper and Clark (1982, pp. 61–2) pointed out:

The treatment of technological change in growth models has . . . an other-worldly degree of abstraction, which is apt to induce considerable scepticism, in spite of its logical elegance. . . . As a result of the traditions in neo-classical growth theory, there is an abiding preoccupation with the possibilities of stable full employment growth with technological change. Since the early post-Keynesian period there has been little work in the Harrodian tradition. A natural extension of Harrod's analysis is to ask whether technological change itself is likely to ameliorate or aggravate instability problems, or to facilitate or work against a correspondence between an actual growth path and the full employment growth path. It is probably fair to say that questions like these, which economists like Harrod would have found quite natural, have been largely side-stepped, at least in the predominant tradition. At one level, this is probably a natural outgrowth of an approach to technological change which starts out from a model in which the destabilising effect of an inequality between planned saving and planned investment is ruled out of consideration from the start. Since this basic source of instability in the real world is left out the question of whether technological change has systematic effects on the stability of economic growth is largely obscured. At another level, the basic assumptions which ensure full employment growth paths in formal growth models, i.e. the possibilities of substituting capital for labour and the existence of perfect factor markets which ensures that this substitution occurs to precisely the extent required, are retained when growth models are extended to take technological change into account.

Among those who have contributed to postwar research on technical change and on the sources, nature and economic consequences of technical innovation, there are some very important differences of method and theoretical approach (Nelson and Winter, 1977, 1982; Rosenberg, 1965,

1982 and 1985; Pavitt 1980 and 1984; Freeman 1982). They nonetheless, to different degrees, all point towards the explanation of some common 'stylized facts', which have been summarized as follows by Dosi (Dosi and Soete, 1983):

First, the innovative process has some *rules of its own* which, at least in the short and medium terms, cannot be described as simple and flexible reactions to changes in market conditions. It is the nature of technologies themselves, which determines the range within which products and processes can adjust to changing economic conditions and the possible directions of technical progress.

Second, scientific inputs play and have been playing a crucial role (increasingly in this century) in opening up new possibilities of major technological advances.

Third, the increasing complexity of research and innovative activities militates in favour of institutional organizations (R & D laboratories of big firms, government labs, universities, etc.) as opposed to individual innovators, as the most conducive environments to the production of innovations.

Fourth, in addition to the previous point, and in many ways complementary to it, a significant amount of innovation and improvements is originated through 'learning by doing', and is generally embodied in people and organizations (primarily firms).

Fifth, notwithstanding the increasing institutional formalization of research, invention and innovation activities maintain an intrinsic *uncertain nature*. The technical (and, even more so, the commercial) outcomes of research activities can hardly be known *ex-ante*.

Sixth, technical change does *not* occur randomly, for two main reasons. First, the *directions* of technical change are often defined by the state-of-the-art of the technologies already in use. Second, it is generally the case that the probability to make technological advances of firms, organizations and often countries, is, among other things, a function of the technological levels already achieved by them (Nelson and Winter 1982 formalize this process in their models through Markovian chains).

Seventh, the evolution of technologies through time presents some significant regularities. When a technology develops and matures it often tends toward (i) exploitations of economies of scale, and (ii) mechanization of production processes.

The implications of these 'stylized facts' for capital–labour substitution are crucial. They tend to support the view of innovative activity as 'strongly *selective, cumulative* in nature and *finalized* in quite precise directions' (Dosi and Soete, 1983). They also lend support to a representation of technological progress in which some new techniques of production may indeed prove superior to others for every income distribution. Thus production techniques embodying microelectronic devices may be superior to equivalent electromechanical ones for every wage and profit rate.

There are two major differences of emphasis which distinguish our approach (and that of other economists who have concentrated on the study of technical innovation), from that of mainstream theorists. First and following on from what was said above, we attach less importance to the 'production function', which figures so prominently in much textbook analysis, whether at the micro or the aggregate level. We argue that these factor-price substitution mechanisms are relatively limited in scope, especially in manufacturing and in the capital intensive 'network services' such as energy and transport, not least because of the nature of technical change itself.

The notion that a temporary reduction in real wages will somehow 'induce' the development and adoption of more labour-intensive techniques in manufacturing is in our view particularly dubious and based on several fundamental fallacies about the nature and direction of technical change. Even in the service sector, as we shall see in chapter 9, recent capital-embodied technical change and the growth of capital-intensity have greatly limited the scope for this type of factor substitution.

This leads on to the second major difference of emphasis between our approach and that of much mainstream analysis. We attach more importance to the process of technical change itself and in particular to major changes in technology affecting many industries and services, and influencing investment and demand throughout the system. Such technological revolutions can save both labour and capital and are the most important source of new investment and employment opportunities. These discontinuous elements in technical change are often ignored by orthodox approaches, which abstract from the specifics of technology and concentrate in our view too much on static analysis of choice of technique and factor substitution.

We believe these differences of emphasis to be of great importance when it comes to analysing the long-term policy implications of technical change and the structural and institutional adaptations involved in the adoption of major new technological systems.

We therefore intend to discuss both of these points, starting with the 'production function' issue in this chapter. In the next chapter we develop further the analysis of major technological transformations in relation to the introduction of microelectronics and other new technologies in the postwar UK economy. In our view this technological transformation offers great possibilities for future employment growth, as well as great productivity gains in most branches of the economy. This preliminary clarification is an essential first step to the detailed discussion of the UK manufacturing and service sectors in chapters 8 and 9.

The 'Production Function' and Factor Substitution

The production function approach to the choice of techniques purports to specify a range of alternative combinations of labour and capital, which

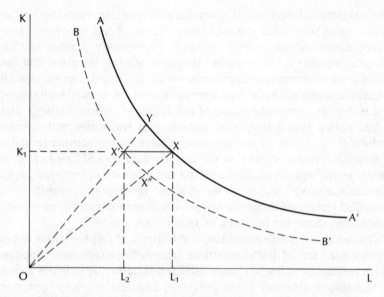

Figure 3.1 The production function

would be capable of producing a given output, within the existing state of knowledge in technology. The particular choice of technique is supposed to depend in any economy, operating under conditions of perfect competition, on the relative prices of labour and capital. When labour becomes relatively more expensive in relation to capital, a more capital-intensive technique should be selected and vice versa. In terms of textbook exposition, the function is usually represented as a smooth curve as in figure 3.1. A change in relative prices is supposed to lead to a movement along the curve, as along AA' in figure 1, while a technological innovation is supposed to lead to a *shift* in the production function to a new position, e.g. from AA' to BB' in figure 3.1.

Technical change is traditionally classified according to its effects on the productivities of the factors of production (normally capital and labour) for a given level of output. Within the neo-classical tradition of a perfectly malleable capital stock, where technical progress is 'disembodied', i.e. unrelated to current investment decisions, the following cases are often distinguished:

  (1) Labour-augmenting (Harrod neutral) technical change, which allows the same output to be produced with less labour for a given capital stock. This type of technical progress acts as a substitute for labour; output can be increased with a given labour input while the capital:output ratio remains constant.

(2) Capital-augmenting (or Solow-neutral) technical change, which correspondingly serves to increase the efficiency of capital.

(3) Hicks neutral technical change, where the efficiencies of both labour and capital are increased such that the 'optimum' capital:labour ratio remains unchanged.

With a continuous production function of the form of figure 3.1, it is not possible to assert a priori which 'type' of technical change is involved in moving from AA' to BB' – indeed, one form of production function (the Cobb–Douglas) is consistent with all three 'types' of technical change. A movement from point X to point X', where the capital:output ratio remains constant, is consistent with Harrod-neutral technical progress. A move from X to X" (with constant capital:labour ratio) is consistent with Hicks-neutral technical progress. But taking each point on AA' in turn, it is clear that a curve like BB' will be obtained whichever point on that curve is regarded as corresponding to any given point on AA'.

Similarly, it is not possible in this context uniquely to describe observed changes in the economy as corresponding to one definition of technical change as opposed to another. Suppose that the economy is initially at point X, with labour input L_1, and capital input K_1. At some later time it may have moved to point X' with employment having fallen to L_2 while the capital stock (and output) are unchanged. This movement could come about through direct labour-augmenting technical change (movement along XX') or through an initial substitution of capital for labour (movement from X to Y), followed by technical progress which is equally labour and capital augmenting (movement along YX'). There are of course an infinite number of other possible routes. Allowance for a continuous spectrum of substitution possibilities means that terms such as 'neutral' or 'labour-saving' technical progress must strictly be defined with respect to assumptions regarding marginal products and returns to production factors, which are supposed to represent marginal products and which implicitly specify the substitution effects incorporated into the definition. For example, technical progress is Harrod-neutral if the capital:output ratio is constant when the rate of profit (or marginal product of capital) is constant, (and is labour saving if a constant rate of profit is accompanied by a rising capital:output ratio).

If we drop the assumption that the factors of production can be combined in an infinite number of alternative proportions to produce a given output, and adopt the opposite limiting case that only one combination of factors is optimal whatever the prevailing relative factor prices, the discussion is greatly simplified. In this case labour-augmenting (Harrod-neutral) technical change is identified unequivocally with a constant capital:output ratio and a declining labour:output ratio, capital-augmenting technical change with a constant labour:output ratio and declining capital: output ratio, and Hicks-neutral technical change with both ratios declining at the same rate. In this case the observation that the capital:output ratio has been increasing (see chapters 7, 8 and 11) implies a labour-saving bias to technical change under any of the definitions; we could say that technical progress has been labour 'augmenting' but capital 'decreasing'. It could be argued that there is effectively a partial substitution process here, brought about by relative factor prices. If labour were free, there would be no incentive for industry to employ techniques involving more capital per unit of output. However, the extent to which a 'substitution' process is involved is, in our opinion, far better regarded as being brought about by the adoption of new technologies rather than by movements along a production function such as AA' in figure 3.1.

This view is illustrated by Atkinson and Stiglitz (1969) who pointed out that the production function is not the smooth curve which is usually portrayed and that shifts in the function usually entail a movement at one particular point, rather than the type of shift illustrated in figure 3.1. An alternative and more realistic representation of many micro level production functions is shown in figure 3.2 (Atkinson and Stiglitz, p. 576). Each point on the original production function (A, B and C) represents a different production process and associated with each one is certain technical knowledge specific to that technique. Since much technique change results from the accumulation of technical knowledge about specific techniques within firms, a change in the production function is often of the type at point A in figure 3.2. In this situation, at least for some time, there is little or no elasticity of substitution between labour and capital in the new

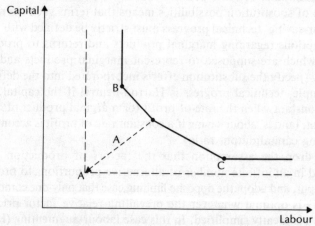

Figure 3.2 The production function.
Source: Atkinson and Stiglitz (1969).

production function (portrayed by the dotted line). They comment that 'in some industries the effect of localization has been so strong that the advanced techniques dominate the less capital-intensive ones, requiring both less labour and less capital'.

So far as the manufacturing industries are concerned we accept that the Atkinson–Stiglitz formulation in figure 3.2 is much more realistic than the traditional representation in figure 3.1. Our own simulation modelling for the manufacturing sectors (see chapter 6) is based on this assumption. Furthermore, as we shall see in chapter 9, there are other industries, such as energy and transport in addition to manufacturing, where in our view it is unrealistic to assume that a reduction in the relative price of labour would in itself lead to the hiring of additional workers. This is not to say that there are not still some service activities where wage reductions could have some employment generating consequences, as a result of increased profitability rather than factor substitution.

The emphasis on factor substitution between labour and capital is also in our view often misleading for many other reasons. Among the principal ones are the following: it ignores the inter-relatedness and complementarities of many technical and organizational innovations, thus tending to assume a spurious 'reversibility' and a wider spectrum of choice in technology, than exists in reality. It disregards the heterogeneity of many production inputs and the specificity of particular skills and types of production equipment. Above all, it ignores the extent to which particular 'technological trajectories' (Dosi, 1983; Nelson and Winter, 1982) dictate the path of technical change for enterprises which aspire to remain competitive in international technological competition. As we have seen, it also ignores the firm-specific nature of much technical innovation and technological accumulation. The individual firm does not start from a blank sheet, but with a set of existing resources, technical capabilities, traditions and limitations (Atkinson and Stiglitz, 1969; Pavitt, 1983).

The neo-classical production function theory resembles the analogous Heckscher–Ohlin theory of factor proportions in international trade in its over-preoccupation with relative prices of labour and capital, rather than technical novelty and good design; it also resembles that theory in its relatively weak empirical validation. Finally, it substitutes a non-existent perfectly informed entrepreneurial rationality, selecting from an imaginary wide spectrum of choices of various factor combinations of labour and capital, for the real uncertainties, rules of thumb, social constraints, technological trajectories and imperatives which actually guide most entrepreneurial decisions in relation to technological choice (Rosenberg, 1975, 1982; Nelson and Winter, 1982; Dosi, 1984).

Factor Prices and 'Induced' Technical Change

This deep-going critique of traditional production function theory should not be taken as a rejection of all inducement mechanisms affecting the direction of technical change, as a result of changing factor prices. On the contrary we fully accept the reality of such mechanisms. Our critique relates to the unrealistic formulation of the ways in which such mechanisms operate and in particular to the oversimplified and schematic way of looking at the issue of labour–capital substitution.

In considering the behaviour of designers, engineers and entrepreneurs in the development and selection of technologies we would stress the importance, on the one hand of persistent long-term trends in relative factor prices (rather than minor short-term fluctuations), and on the other hand, like Rosenberg (1975) we would also stress the influence of rather sudden, dramatic changes, such as the oil price shock or the effect of shortages on materials prices.

Since the long-term expectation of engineers and designers is that the price of labour will slowly increase, relative to that of capital, it is unlikely that short-term fluctuations would reverse this general expectation.

Salter was overstating his case when he suggested that factor prices do not 'induce' innovation in particular directions. Even where innovation is 'exogenous', cost considerations in relation to particular inputs are often present in the minds of scientists and engineers. Thus, for example, the two oil 'shocks' of 1973 and 1979 undoubtedly 'induced' both investment decisions in favour of less energy-intensive systems within the bounds of existing technology and innovative effort directed specifically to new energy-saving techniques. The range of techniques 'on offer' has thus certainly been influenced by the major changes in the price of energy, and the same applies to other big shifts in relative prices, which are expected to persist.

This case also illustrates rather well our point about technological trajectories. It took several years for the effects of the oil shocks to work their way through the system and it may be that the '1979 shock' had a greater effect than the '1973 shock'. The first effects were of 'retro-fitting' energy-saving devices to existing plant and heating systems. (This could be regarded as movement 'along' the production function within the bounds of existing technology.) It was only some time later that new plant and machinery began to be installed embodying technology with greater energy-saving potential. It is highly unlikely that this trajectory will be reversed easily by the more recent fall in energy prices, unless this proves to be both very large and persistent, because decision-makers are guided by what they believe to be major and long-term trends, not by what they

Table 3.1 Energy and economic growth, IEA member countries, 1968–83

	1968	1973	1978	1983	Annual average change (%)			
					1968–73	1973–8	1978–83	
Gross domestic product (GDP)[a]	3177	3713	4227	4595	3.2	2.6	1.7	
Total primary energy requirements (TPER)[b]	2586	3324	3592	3359	4.1	1.6	-1.3	
Total final consumption (TFC)[b]	1929	2491	2609	2398	4.2	0.9	-1.7	
TPER/GDP[c]	0.81	0.90	0.85	0.73	2.1	-1.1	-3.0	
TFC/GDP[c]	0.61	0.67	0.62	0.52	1.9	-1.5	-3.5	

[a] Billion US$, 1975 prices and exchange rates.
[b] Million tonnes of oil equivalent; total final consumption equals total primary energy requirement less transformation and distribution losses.
[c] Tonnes of oil equivalent per thousand US$.

Sources: Energy Policies and Programmes of IEA Countries, 1983 Review, International Energy Agency, Paris, 1984; *Energy Balances of OECD Countries, 1960–74*, OECD, Paris, 1970: *Main Economic Indicators*, OECD, Paris, various issues; W. Walker, Information Technology and the Use of Energy, *Energy Policy*, October 1985.

perceive as short-term fluctuations. Once set in motion these trajectories are not easily reversed (table 3.1).

We shall argue in chapter 4 that the drastic reduction in costs of processing, storing and transmitting information, as a result of the microelectronic revolution, has also had profound effects on design, development and investment decisions throughout the economic system, because engineers, designers and entrepreneurs expect them to persist.

We would stress long-term persistence, because we agree with Atkinson and Stiglitz (1969) that decisions about technology are not simply taken on the basis of current prices. As they forcefully point out (p. 575): 'A firm will not necessarily allocate research expenditure to the technique in current use when it expects that rising wages will lead it to develop and use a more capital-intensive technique in the future. . . . the firm cannot behave myopically.'

An important conclusion follows from this overall assessment of 'induced innovation'. It is, that there is inherent plausibility in the Hicks inducement theory, biasing the long-term direction of technical change in a labour-saving direction. Attempts to generate a reversal of this trend by temporary small reductions in the relative price of labour are extremely unlikely to be effective. Only a reversal of long-term expectations and/or a sudden drastic reduction in real wages comparable to the oil 'shocks' would be likely to prove effective and in our view the social consequences of such a change could prove disastrous.

This does not mean of course that it does not make sense to develop social mechanisms to restrain the rate of growth of real and nominal incomes in line with productivity growth to prevent inflationary pressures, or even that local labour markets may not benefit from some temporary downward flexibility. Our critique refers to a systematic attempt to bring about substitution of labour for capital by overall reductions in real wages. The argument about profitability is distinct from the labour–capital substitution argument.

The 'factor substitution' approach in production function analysis tends to focus attention on comparative statics. But from a dynamic standpoint it is the shifts in production function which matter more than movements along it. These often depend on cost changes affecting other production inputs as well as labour. Even more crucial this approach tends to ignore product innovation, service innovation and system innovation with entirely new production functions (see, for example, Stoneman, 1983, p. 170). When we are considering the long-term growth of new employment, then the role of new activities, industries and services must be a central concern.

Again in their critique of mainstream formulations of the neo-classical production function, Atkinson and Stiglitz (1969, p. 577) pointed to some of these fundamental weaknesses:

. . . at least two unreasonable assumptions are employed . . . in the Drandakis-Phelps (1966) and Samuelson (1965) models the choice of the kind of factor-augmenting technical progress is made on the basis of current factor prices only. More fundamentally, the firm's choice is restricted to purely factor-augmenting technical progress. Would it really want to raise productivity on handcarts as well as forklift trucks?

In Third World countries in protected industries or activities, such as road construction, or some service activities where international competition is not a major consideration, the choice of 'pick and shovel' techniques employing a great deal of low cost labour may be an economically 'rational' choice and contribute to employment goals. It may also make sense to concentrate some R & D effort on labour-intensive techniques or on 'handcarts' rather than forklift trucks or automated warehouse systems. But for a country such as the United Kingdom, which is committed to compete in open international markets on a long-term basis, the choice of internationally competitive technology is crucial. This means that employment generation must mainly be sought by the widespread adoption of advanced technologies and not by the transition to a low wage, technologically backward service economy. To rely on low wages as an inducement for the choice of labour-intensive techniques and the major source of competitive leverage in world markets would be a path of voluntary underdevelopment.

In dynamic terms the choice of capital-intensive techniques, even in Third World countries with very low labour costs, is frequently not so 'irrational' as it appears, as the choice of a labour-intensive alternative would often mean a decision to opt for a technologically backward solution away from the mainstream of technical advance, which would rule out the possibility of an internationally competitive industry.

In relation to the issue of employment generation, all of this is not to dispute the existence of long-term compensation mechanisms, based on income and price elasticities. But whereas most neo-classical theory on technical change puts the main emphasis on factor substitution and hence on price elasticity and process innovation, we would stress far more the over-riding importance of income elasticity of demand in relation to new and improved products and the rise of new industries. Katsoulacos (1985) has shown, even within the terms of neo-classical theory, that the employment benefits of product innovations are greater and more immediate than those of process innovations, provided the new industries and services are established within the country under consideration.

As Jones (1980, p. 27) pointed out:

Technological unemployment did not appear to be a problem in the 1950s and 1960s not because the technology was absent or its diffusion slow. Technological unemployment was absent because sufficient compensatory mechanisms were generated by the technology itself and by the growth of the economic system as a whole.

Paradoxically, whereas a traditional neo-classical analysis would present today a relatively bleak empirical assessment of the employment compensation potential of new technologies given the low price and substitution elasticities in most industries, our own analysis would come up with a more optimistic compensation scenario pointing in the first instance to the employment growth opportunities in the new technology industries and products both in terms of the domestic market and in terms of the export market. However, as we shall show in part IV the potential for compensation would be crucially dependent on the supply of capital and new investments, and on the development of an appropriate institutional framework to give the new technologies full scope and to retain a competitive position in international trade.

4
Information Technology and Change in Techno-Economic Paradigm

Christopher Freeman

When we are considering which technologies are likely to be the most important for the future development of the United Kingdom and other industrial economies, there would probably now be fairly widespread agreement that 'information technology' is the key. We are using the expression 'information technology' (IT) to describe a combination of new technology developments in microelectronics, computers, optoelectronics and communication systems, which collectively are having such pervasive effects that many people have talked in terms of a 'paradigm' change in technology.

Much of the recent debate on technical change and employment has focused on the effects of 'microelectronics' or 'automation' and on the labour-displacing impact of process innovations, such as computer-controlled machine tools, robotics or word processors. For reasons which will have become clear in chapter 3 we prefer to use the wider expression 'IT', since we are concerned with the employment generating effects of new products and new systems, as well as with the job-displacing effects of various types of 'automation' and other process innovations.

In the course of our research project we became convinced that IT is indeed an all pervasive change, which is affecting the design of many existing products and services as well as the mode of producing and marketing almost all of them. This point could be illustrated from any of our sector studies and is summarized in chapter 10. It is confirmed both by technological evidence such as the general trends in R & D, patenting or innovation and by economic evidence on investment, output and trade.

49

Four Approaches to Information Technology

In this section we shall examine four different approaches to the problem of defining the scope of the new technologies and measuring their employment effects.

The first approach regards the debate about new technology as essentially a continuation of the 'automation' debate of the fifties and uses the expressions 'factory automation' or 'office automation' as though they were virtually synonymous with information technology. Those who take this view are sometimes rather impatient with the contemporary discussion since they regard all the important questions as having been resolved long ago by economists such as Einzig (1957) or others who contributed to the report of the *US National Commission on Technology, Automation and Economic Progress* (1966). The emphasis in this approach is almost entirely on process innovation.

The second approach regards the IT 'industries' as major new branches of the economy capable of imparting an upward impetus to employment in their own right. Thus, for example, it is pointed out that as a result of its extraordinary growth in the sixties and seventies, the computer industry in the United States now employs more people than the automobile industry, once regarded as a typical 'engine of growth' of employment and economic activity more generally. In 1982 in the United States the computer industry (3573) employed 340,000 people whilst the motor vehicle industry (3711) employed 239,000 people (in both cases excluding components) (US Industrial Outlook, 1985, Department of Commerce). The 'IT sector' in this approach comprises both manufacturing and service industries, whose growth may be analysed in the same way as that of vehicles, electrical machinery or garages and motor repair services, once the necessary reclassification of industrial output and employment statistics has been satisfactorily performed.

The third approach, pioneered by social scientists such as Daniel Bell (1974) and Fritz Machlup (1962), puts the emphasis on information activities wheresoever they are performed. Machlup's pioneering book was entitled *The Production and Distribution of Knowledge*, whilst Bell introduced the notion of the 'postindustrial society'. Sociologists and economists taking this route tend to stress the growth of information-related or 'knowledge producing' occupations in every industry and service as a long-term trend characteristic of the twentieth century, and leading to what is often described as the 'information society'. This approach is not necessarily concerned mainly or exclusively with electronic or computer technology, although these have received increasing emphasis in recent times.

The fourth approach, which we shall ourselves adopt, comprises elements of all three of those which have been described and defines 'information technology' both as a new range of products and services, and as a technology which is capable of revolutionizing the processes of production and delivery of all other industries and services. The scope for such a new technology has been vastly enlarged as a result of the long-term trends identified by the third approach, but the technology itself is new, having emerged in the last couple of decades as a result of the convergence of a number of inter-related radical advances in the field of microelectronics, fibre optics, software engineering, communications and computer technology. An approach to information activities which ignores the specific features of the new technologies is in danger of overlooking many of the economic and social consequences of these technologies, including their employment effects. Our approach puts the emphasis on the new technology and not just on the information. However, in chapter 10 we do make use of the concept of an 'IT sector' comprising both the 'IT producing sector' (as used in the second approach outlined above) and an 'IT-user sector' composed of those industries and services which are making the most intensive use of IT. Ultimately the 'IT-user sector' could grow to comprise many other sectors of the economy, so that our 'IT sector' approach serves to illustrate the process of structural change.

We shall argue that only this fourth approach can yield satisfactory results from the standpoint of analysing the overall economic and employment effects of IT, since the first approach (automation) has an implicit bias towards job displacement, and the second (IT industries) towards job creation, and each is needed to complement the other. The third approach is more a theory of occupational trends than of employment. But the new information technologies affect industrial structure as well as occupational structure. Moreover, they have other specific features which cannot be ignored in considering the problems of structural change and adjustment which have arisen in all countries.

Limitations of Three Approaches to Information Technology

We now briefly consider each of these four approaches in order to demonstrate their relevance to the contemporary debates. We first consider the merits and shortcomings of looking at 'information technology' simply or mainly as 'automation'.

The earlier debates on 'automation', particularly in the United States and in the fora of international organizations, foreshadowed in some respects the present debates on the employment effects of information technology. However, it is important to recognize some of the differences as well as the similarities between the two debates. Since the fifties the technologies involved have advanced enormously; at the same time the economic and social climate has also changed, together with the potential areas of application.

This emerges very clearly, for example, if we examine one of the major contributions by economists, Paul Einzig's *Economic Consequences of Automation* (1957, pp. 2–3). Whilst Einzig covers much of the ground which is still debated, and refutes a number of fallacies which are still prevalent, his discussion is 'dated' in several important respects.

In the first place it is clear that although he opts for a very broad definition of 'automation', he sees it primarily as a very advanced form of mechanization and as concerned fundamentally with saving labour:

Fortunately for the writer of this book, and for the majority of his readers, there is no need, from the point of view with which we are here concerned, to choose between the multitude of existing definitions, or to produce an original definition, in order to determine the scope of the subject too rigidly. Such a definition or an enumeration and description of the whole wide range of processes that come under it would be irrelevant from an economic point of view. What matters for us is that automation, whether in the broadest or the narrowest sense of the term, is a technological method that tends to reduce current production costs in terms of man-hours per unit of output. . . . Its loose use as a synonym for advanced mechanisation may shock the technologist, but serves the purpose of the economist.

Although Einzig does acknowledge that automation in its broadest sense can bring other advantages besides saving labour, such as the provision of better and quicker information, most of the examples that he gives are of 'advanced mechanization'. Moreover, he sees automation as a continuation of the trend towards dedicated mass and flow production systems, turning out standardized products. He does not discuss telecommunications or the role of new types of information services, nor does he mention semiconductor technology. Although he expected the costs of automation equipment to fall, he could not have foreseen the drastic reduction in costs of storing, processing and transmitting information which occurred in the sixties and seventies, as a result of revolutionary advances in microelectronics, in optoelectronics and in related fields.

Again, although he mentions 'office automation' and gives one example of the applications of computers in a mail-order firm, his emphasis is almost entirely on applications in manufacturing industry and there is little or no discussion of future developments in service industries.

These points are made not in order to belittle the work of Einzig – his book, now almost 30 years old, remains one of the best discussions of the economics of automation – but in order to bring out the point that the concepts of 'automation' often used in the fifties are inadequate and indeed misleading in relation to the understanding of contemporary problems. Even at the time technologists and industrialists, such as Leon Bagrit (1965, pp. 19–20), pointed out the very unsatisfactory and limited view of 'automation' which was implied in much discussion of the fifties:

I have always thought it was most unfortunate that the word 'automation' was invented in the motor car industry, in fact by a works manager at Ford's in Detroit. . . . The word . . . has unfortunately developed the connotation not of cybernation, but of automaticity and there is the greatest difficulty of getting rid of this unfortunate meaning.

We prefer to talk about 'information technology' rather than 'automation' for this and other reasons, but most of all because it reflects more accurately the convergence of computer technology with related developments in telecommunications and in microelectronics. This combination is extremely powerful, in terms both of the economic and of the technical advantages, and it means that the actual and potential applications of the new technology go far beyond those which were customarily considered in the fifties debates on automation, even though they were the subject of speculation by some of the more imaginative scientists and technologists.

A second reason for preferring the concept of 'information technology' rather than 'automation' is fundamental to the consideration of the employment issue. 'Automation', as we have seen in the case of Einzig, tends to carry with it, even when it is broadly defined, the connotation of an exclusive concern with process innovation. Although labour-saving technical change is most certainly one of the major economic advantages of the new technology, this aspect has been greatly overemphasized at the expense of other equally important characteristics.

Thus, information technology has already demonstrated very great advantages in material-saving, energy-saving and capital-saving applications, as for example, in the reduction of the number of mechanical and electronic components in a wide range of engineering products from cash registers to machine tools, or in the reduction of stocks of components, materials, work-in-progress, and finished products, through far more efficient systems of inventory control and better communications with suppliers and distributors alike (see Perez, 1985).

More fundamentally, the emphasis on labour-saving process automation technology tends to overshadow the development of new products and new services associated with the new technology. Some of the more pessimistic forecasts of future employment prospects are based on the view that job displacement through automation will tend permanently to outstrip new job creation. It is extremely important, as is also evident from the theoretical discussion in chapter 2, to look both at job creation and at job displacement.

In the second approach discussed above, a group of industries and services are identified and classified as the 'IT sector', as for example in the recent *OTA Report* (1985). Usually the IT industries are defined as a grouping around the computer industry, comprising also electronic components, other electronic capital goods and communication equipment. In this definition they are almost identical with the electronics industries, which have indeed been almost the only growth sector within manufacturing in terms of employment, in several OECD countries over the past decade. In addition to the IT industries in the manufacturing sector, it is also possible to identify a range of IT service industries, such as software engineering, business computer consultancy, broadcasting, and other information services. Some of these areas have seen the most dynamic employment growth of any in the past 10 years, although they are rarely satisfactorily identified yet in national and international output, investment and employment statistics. We return to this issue in chapter 10, where we stress the importance of considering also the major IT-user industries, which are playing an increasingly active and important role in the further development of the technology.

We now turn to consider the third approach based on the efforts by sociologists, such as Daniel Bell (1974), and economists, such as Machlup (1962) to demonstrate that modern industrialized societies are becoming 'information societies'. Although the concern with the importance of 'information' activities is common to these theorists and to our own purpose in this monograph, there is an important difference of emphasis, definition and classification.

Machlup, Bell, and Porat (1976) were primarily concerned not with technology, but with demonstrating that the social and occupational composition of the work force, and of the population in general, was shifting from 'blue collar' jobs to 'white collar' jobs. They associated this change with the growth of all kinds of clerical, technical, managerial, educational and related occupations. Machlup enumerated all those occupations which were 'producing or distributing knowledge' rather than materials or goods, and estimated that even in 1959, they accounted for nearly a quarter of the total work force in the United States. More recently, Porat (on somewhat broader definitions) estimated the proportion of information workers in 1974 as half the total labour force.

Although this change in the occupational and social composition of the population is an important factor enlarging the scope of information technology as defined here, it is a process which is distinct from those technical and organizational changes that are our central concern. The concept of 'information technology' which we are using is based on a taxonomy of technology rather than a taxonomy of occupational structure. Information technology is comparable to the introduction of electric power in its capacity to transform other branches of the economy, as well as to impel the growth of entirely new branches.

A Taxonomy of Technical Change

In order to illustrate the nature of IT and its significance for the economics of employment, in our own approach it is necessary to categorize various types of technical change and to distinguish between:

Incremental Innovations

These occur more or less continuously in any industry or service activity, although at a varying rate in different industries and over different time periods. They may often occur, not so much as the result of formal research and development activity, but as the outcome of inventions and improvements suggested by engineers and others directly engaged in the production process, or as a result of initiatives and proposals by users. Many empirical studies have confirmed their great importance in improving the efficiency in use of all factors of production, for example, Hollander's study (1965) of productivity gains in Du Pont rayon plants or Townsend's (1976) study of the Anderton shearer loader in the British coalmining industry. They are particularly important in the follow-through period after a radical breakthrough innovation (see below) and frequently associated with the scaling up of plant and equipment and quality improvements to products and services for a variety of specific applications. Although their combined effect is extremely important in the growth of productivity, no single incremental innovation has dramatic effects, and they may sometimes pass unnoticed and unrecorded. However, their effects are apparent in the steady growth of productivity which is reflected in input–output tables over time by major changes in the coefficients for the existing array of products and services.

Radical Innovations

These are discontinuous events and in recent times are usually the result of a deliberate research and development activity in enterprises and/or in university and government laboratories. They are unevenly distributed over sectors and over time, but our research did not support the view of Mensch (1975) that their appearance is concentrated particularly in periods of deep recessions (Freeman et al., 1982). But we would agree with Mensch that, whenever they may occur, they are important as the potential springboard for the growth of new markets, or in the case of radical process innovations, such as the oxygen steelmaking process, of big improvements in the cost and quality of existing products. Over a period of decades a radical innovation, such as nylon or the contraceptive pill, may have fairly dramatic effects, but in terms of their economic impact they are relatively small and localized, unless a whole cluster of radical innovations are linked together in the rise of entire new industries and services, such as the synthetic materials industry or the semiconductor industry. Strictly speaking, at a sufficiently disaggregative level, radical innovations would constantly require the addition of new rows and columns in an input–output table. But in practical terms, such changes are introduced only in the case of the most important innovations and with long time-lags, when their economic impact is already substantial.

New Technological Systems

Keirstead (1948), in his exposition of a Schumpeterian theory of economic development, introduced the concept of 'constellations' of innovations, which were technically and economically inter-related. Obvious examples are the clusters of synthetic materials innovations and petrochemical innovations in the thirties, forties and fifties. We described these clusters as 'new technological systems' in our earlier book (Freeman et al., 1982). They include numerous radical and incremental innovations in both products and processes.

Changes of 'Techno-Economic Paradigm' (Technological Revolutions)

These are far-reaching and pervasive changes in technology, affecting many (or even all) branches of the economy, as well as giving rise to entirely new sectors. Examples given by Schumpeter were the steam engine and electric power. Characteristic of this type of technical change is that it affects the input cost structure and the conditions of production and distribution for almost every branch of the economy. A change in techno-economic paradigm thus comprises clusters of radical and incremental innovations and embraces several 'new technology systems'. Once a new techno-economic paradigm has become established throughout the economy it may be described as a 'technological regime'.

Nelson and Winter (1977) have also used the concepts of 'technological regimes' and of 'natural trajectories' in technology. Their 'general natural trajectories' correspond perhaps most closely to our 'paradigms'.

Dosi (1983) has used the expression 'change of technological paradigm' and made comparisons with the analogous approach of Kuhn (1962) to 'scientific revolutions' and paradigm changes in basic science. In these terms 'incremental innovation' along established technological trajectories may be compared with Kuhn's 'normal science'. Whilst strict analogies are out of place the concept of 'paradigm' change has the merit of bringing out the element of inertia in the system.

Whilst there are similarities in all these concepts, the approach of Perez (1983, 1984) is the most systematic and has some important distinguishing features in relation to the structural crises of adaptation with which we are concerned. She argues that the development of a new 'techno-economic paradigm' involves a new 'best practice' set of rules and customs for designers, engineers, entrepreneurs and managers, which differs in many important respects from the previously prevailing paradigm. In terms of the previous discussion on production functions, such technological revolutions give rise to a whole series of rapidly changing production functions for both old and new products. Whilst the exact savings in either labour or capital cannot be precisely foreseen, the general economic and technical advantages to be derived from the application of the new technology in product and process design become increasingly apparent and new 'rules of thumb' are gradually established. Such changes in paradigm make possible a 'quantum leap' in potential productivity, which, however, is at first only realized in a few leading sectors. It takes decades for the productivity gains to be realized throughout the economy as a result of a process of learning, adaptation, incremental innovation and institutional change.

Changes of techno-economic paradigm are based on combinations of radical product, process and organizational innovations. They occur relatively seldom (perhaps twice in a century) but when they do occur they necessitate changes in the institutional and social framework, as well as in most enterprises if their potential is to be fully exploited. They give rise to major changes in the organizational structure of firms, the skill mix and the management style of industry.

The overwhelming importance of such technological transformations is that, if the problems of institutional adaptation can be overcome, they offer tremendous scope for new employment-generating investment as well as labour-saving productivity gains. These opportunities arise both in the provision of new and improved consumer goods and services, and in the provision of a new range of capital equipment for all sectors of the economy. Characteristically, they facilitate savings in all factors of production, although initially their effects may be concentrated on one particular factor.

Perez (1983) has suggested that big boom periods of expansion occur when there is a 'good match' between a new techno-economic 'paradigm' or 'style' and the socio-institutional climate. Depressions, in her view, represent periods of mismatch between the emerging new paradigms (already quite well advanced during a previous long wave of expansion) and the institutional framework. The widespread generalization of the new paradigms, not only in the 'leading' branches of the upswing but also in many other branches of the economy, is possible only after a period of change and adaptation of many social institutions to the requirements of the new technology. Whereas technological change is often very rapid, there is usually a great deal of inertia in social institutions, buttressed by the political power of established interest groups, as well as by slow response times of many individuals and groups.

Schumpeter emphasized the importance both of organizational innovations and of technical innovations, and of their interdependence. This combination is a major characteristic of a change of technological paradigm, such as the introduction of information technology, and leads to changes in management structure as well as process technology. An example of these changes is given in the article by Gooding (1984) on the 'Cultural Revolution in GM' which describes the major reorganization of one of the world's largest companies as a result of the widespread application of IT in its products and processes. These changes did not simply involve the large-scale introduction of robotics, although this is certainly important, but also included big changes in the organization of design, production and model changes, and the acquisition of several software firms.

An interesting historical analogy to the case of information technology is provided by the experience of electric power. Figure 4.1 illustrates the point that, although Mensch (1975) is correct in drawing attention to a cluster of radical innovations in the 1880s, the major economic effects of electrification came much later with the growth in the share of electricity in mechanical drive for industry from 5% in 1900 to 53% in 1920. This was possible only after the acceptance of a major change in factory organization from the old system based on one large steam engine driving a large number of shafts through a complex system of belts and pulleys, to a system based first of all on electric group drive and later on unit drive, i.e. one electric motor for each machine. Under the old system all the shafts and countershafts rotated continuously no matter how many machines were actually in use. A breakdown involved the whole factory.

An excellent article by Devine (1983, p. 357) 'From shafts to wires' has documented this change in some detail. Devine points out that:

Replacing a steam engine with one or more electric motors, leaving the power distribution system unchanged, appears to have been the usual juxtaposition of a new technology upon the framework of an old one. . . . Shaft and belt power distribution systems were in place, and manufacturers were familiar with their

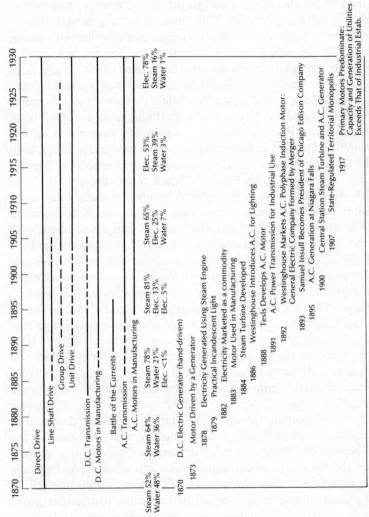

Figure 4.1 Chronology of electrification of industry: (a) methods of driving machinery; (b) rise of alternating current; (c) share of power mechanical drive provided by steam, water, electricity; and (d) key technical and entrepreneurial developments. *Source:* Warren Devine, 1983; From shafts to wires, *Journal of Economic History*, 43, No. 2.

problems. Turning line shafts with motors was an improvement that required modifying only the front end of the system. . . . As long as the electric motors were simply used in place of steam engines to turn long line shafts the shortcomings of mechanical power distribution systems remained.

It was not until after 1900 that manufacturers generally began to realize that the indirect benefits of using unit electric drives were far greater than the direct energy saving benefits. Unit drive gave far greater flexibility in factory layout since machines were no longer placed in line with shafts, making possible big capital savings in floor space. For example, the US Government Printing Office was able to add 40 presses in the same floor space. Unit drive meant that trolleys and overhead cranes could be used on a large scale unobstructed by shafts, countershafts and belts. Portable power tools increased even further the flexibility and adaptability of production systems. Factories could be made much cleaner and lighter, which was very important in industries such as textiles and printing, both for working conditions and for product quality and process efficiency. Production capacity could be expanded much more easily.

The full expansionary and employment benefits of electric power on the economy depended, therefore, not only on a few key innovations in the 1880s or on an 'electricity industry', but on the development of a new 'paradigm' or production and design philosophy. This involved the redesign of machine tools, the introduction of portable power tools and much other production equipment. It also involved the relocation of many plants and industries, based on the new freedom conferred by electric power transmission and local generating capacity. Finally, the revolution affected not only capital goods but a whole range of consumer goods, as a series of radical innovations led to the universal availability of a wide range of electric domestic appliances going far beyond the original domestic lighting systems of the 1880s. Ultimately, therefore, the impetus to economic development from electricity affected almost the entire range of goods and services.

The lags and discontinuities in the adaptation of the social system to the potential of a revolutionary new technology apply not only to the sphere of production. There are similar problems in the sphere of consumer behaviour and the design of consumer goods and services. As Pasinetti (1981) has pointed out, structural crises of adjustment involve not simply marginal changes in consumer expenditure for one or two products, but a new type of consumer behaviour and an entire new pattern of demand. At first the tendency on the supply side is to try and adapt the new types of goods and services to the existing pattern of consumer behaviour.

Once again, the analogy with electric power is a valuable one. As Adrian Forty (1981) describes it in his book on *Objects of desire*:

On the whole, as electricity supply engineers frequently complained, the standard of design was not high. In some cases, such as certain domestic clothes washing machines, the appliances were manual models to which an electric motor had been attached. Some electric cookers had been made by installing electric elements in the cast-iron carcase of gas cookers, while those that were purpose-built often looked very much like gas-cookers. Even when, as in the case of electric fires, the product was not a direct substitute for a manual or gas appliance, there was rarely much about the designs to indicate their specifically electrical nature, and the level of efficiency was generally unspectacular. The extravagant claims that were sometimes made for the luxury of the all-electric home were hardly justified by the appliances themselves, which could not in any case be put to more than occasional use because of the high price of electricity.

The diffusion of new information services, 'intelligent' domestic appliances and systems, and entertainment is still at a comparable level so far as domestic consumers are concerned. It is as hard for us to imagine the future constellation of demand as it was in the 1880s to imagine the typical household with a full range of electrical appliances bought on hire-purchase. Moreover, just as the development of this new pattern of demand for a wider variety of consumer durables and appliances was dependent on massive infrastructural investment, ultimately making available cheap electricity to every household, so today the market for many new services depends on the development of the telecommunications infrastructure. The partial saturation of demand for the old pattern of goods combined with the lag in the development of new services and products designed by consumers is one of the outstanding factors in the process of structural change and growth of new employment.

We would argue that, as in the case of electricity, the full economic and social consequences (including employment generation) of information technology depend on a similar process of social experimentation and learning in exploiting the new technological trajectories. The organizational, social and system innovations at the point of application are just as important as in the case of the electric power industry.

When the universal availability of electricity was followed by very low cost cheap oil, this led to an energy-intensive system of mass production. But, in the present period of structural change, when energy costs have risen and the growth potential of the old leading sectors has been partially exhausted, a new information-intensive 'techno-economic paradigm' has emerged, based on the extraordinarily low costs of storing, processing and communicating information. Because this technology permits far more precise control of all production processes in real time, and of stocks and distribution, it offers great potentiality for cost-savings not just in labour but also in capital, energy and materials. This is linked to a potential for providing a wide range of new products and services. But its diffusion is hindered and its productivity potential is not realized because of institutions, management styles and market structures which are still geared to the old paradigm. The structural crisis of the eighties is in this perspective a prolonged period of social adaptation to this new paradigm.

The new IT paradigm, based on a constellation of industries, which are among the fastest growing in all the leading industrial countries, such as computers, electronic components and telecommunications, has already resulted in a drastic fall in costs and a counterinflationary trend in prices in these sectors as well as vastly improved technical performance. This combination is relatively rare in the history of technology and it means that this new technological paradigm satisfies all the requirements for a Schumpeterian revolution in the economy. Our research showed that this technological revolution is now affecting, although very unevenly, all other sectors.

Thus, our concept of 'information technology' is that it represents both a revolutionary new range of process technologies for all other branches of the economy and a new set of firms and industries, which supply a new range of products and services, including the equipment and software needed to transform the processes of production in the rest of the economy. New types of management structure and business organization accompany these changes. These frequently involve a combination of both 'hardware' and 'software', so that in many cases the boundary-line between a 'manufacturing' firm and a 'service' firm is increasingly difficult to draw.

Characteristics of the 'IT' Techno-Economic Paradigm

We now turn to consider various aspects of the IT 'techno-economic paradigm' as it affects future employment prospects. The reader is referred to various volumes of the sector studies in the Gower Series for a more detailed account of these changes (p. 274). Among the important characteristics of the new paradigm are the following:

(1) A continuing very high rate of technical change in the IT industries themselves, as well as in a wide range of applications. Underlying this process is the continuing dramatic improvement in large scale integration of electronic circuits, and the continuing fall in costs which this permits. As a result of this and parallel developments in optoelectronics and communications technology as well as in computer design and performance, there is a high rate of product obsolescence with rapidly succeeding 'generations' of components incorporated into new designs of end-products and systems. The revolutionary developments in integrated circuits have

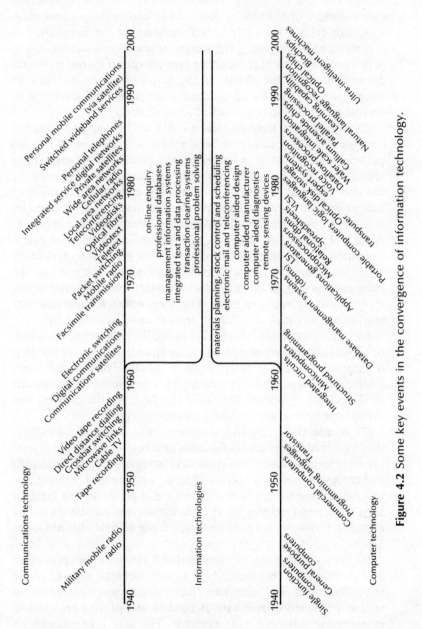

Figure 4.2 Some key events in the convergence of information technology.

Communications technology

1940 — Military mobile radio / radio
1950 — Tape recording / Cable TV / Microwave links / Crossbar switching / Direct distance dialling / Video tape recording
1960 — Communications satellites / Digital communications / Electronic switching / Facsimile transmission
1970 — Mobile radio / Packet switching / Teletext / Videotext / Optical fibre / Videodisks / Teleconferencing / Local area networks / Cellular radio
1980 — Wide area networks / Private networks / Integrated service digital networks / Personal telephones
1990 — Switched wideband services
2000 — Personal mobile communications (via satellite)

Information technologies

professional problem solving
transaction clearing systems
integrated text and data processing
management information systems
professional databases
on-line enquiry
remote sensing devices
computer aided diagnostics
computer aided manufacturer
computer aided design
electronic mail and teleconferencing
materials planning, stock control and scheduling

Computer technology

1940 — Single function computers / General purpose computers
1950 — Programming languages / Commercial computers / Transistor / Programming languages
1960 — Structured programming / Minicomputers / Integrated circuits
1970 — Database management systems (dbms) / Applications generators / LSI / Microprocessors / Relational dbms / Spreadsheets / VLSI
1980 — languages / expert systems / Voice recognition / Optical disk storage / Wafer scale integration / Datalow processors / Gallium arsenide chips
1990 — Natural language recognition / Learning processing / Optical chips / Ultra-intelligent machines / Biochips
2000 — Portable computer (transputer)

their parallel in communication technology, where fibre optics have made possible similar drastic improvements in costs and performance (see volume 3 of the Gower Series). The convergence of all these technological advances (figure 4.2) means that the capability for communicating, processing and storing information is still improving very rapidly and that it is becoming even cheaper. This has profound consequences for the ability of the new technology in terms of integration and control of productive activities, and gives rise to several other characteristics of IT, as follows.

(2) Much greater flexibility and speed in model changes and design changes. Whereas mass and flow production systems of the old paradigm were based on 'dedicated' equipment continuously replicating vast numbers of standardized, homogeneous products, the new flexibility conferred by IT permits more rapid changes of tooling and dies, so that small production runs become economic, and the prospects for small and medium-sized firms are changed. Piore and Sabel (1984) have argued that the availability of cheap computers will change the prospects of flexible small-scale craft-based firms so dramatically as to permit them to predominate in the economy generally, as in the nineteenth century, provided appropriate institutional changes are made. Our own research suggested that marketing, financial and design economies of scale are still very important in most manufacturing industries. Thus the rise of many new small firms is unlikely completely to displace the dominant position of large information-intensive firms. It is more likely to lead a new symbiosis of large firms surrounded by a penumbra of specialized satellite firms. However, there will also be a proliferation of new 'niche' markets served by small specialized firms.

(3) A reduction in electromechanical components and in various stages of component transformation as a result of the redesign of products and processes. This leads to substantial savings in materials and energy, as well as a process of structural change involving loss of jobs in some sectors of metals and metal goods and an increase in jobs in some sectors of the electronics component industry, electronic products, and producer services (see volume 4 of the Gower Series).

(4) As a result of the changes outlined above, and in particular as a result of the continuing high rate of technical change in the microelectronics and computer industries themselves, a strong tendency to a more rapid rate of product and process change and more intense technological competition. This in turn strengthens the demand for new skills and services in design, development, software engineering and IT products generally, affecting both in-house skill

and employment profiles and the growth of new externally-contracted services, information flows and supplies of equipment. The more rapid rate of product and process change is facilitated and stimulated by the diffusion of computer-aided design systems. CAD itself permits the much closer integration of design, production and procurement, and provokes the reorganization of management systems to exploit this possibility (see especially volumes 3 and 5 of the Gower Series).

(5) Speed, reliability and low cost of communicating and storing vast quantities of information relating to sales, inventories and financial transactions generally. Ultimately the whole structure of banking, financial services and distribution is likely to be transformed as a result of the introduction of 'electronic funds transfer' based on this capability. Already the information technology revolution has had very far-reaching effects on banks, insurance companies, retail chains and supermarkets, making possible capital and materials saving as well as labour-saving organizational and technical changes (see volume 5 of the Gower Series).

(6) The capability which IT confers to integrate design, manufacture, procurement, sales, administration and technical service in any enterprise. The ultimate tendency is towards computer-integrated manufacturing systems and all-electronic office systems, but these are still some way off and there are many social as well as technical barriers to their full realization. Nevertheless many enterprises are moving in this direction, as exemplified in the rapid take-off of CAD systems in many branches of manufacturing and their increasingly close linkages with ordering systems for tooling and parts, as well as with manufacturing schedules (see volume 4 of the Gower Series). These developments have already had significant effects on the skill profile of industry, affecting drawing offices and design, as well as clerical labour and machine shops. In the service industries, as well as in manufacturing, the skill profile and the levels of employment in offices are being affected by the introduction of word processors and their integration with new communication systems and information services. Whilst these new developments undoubtedly displace some clerical labour, some draughtsmen and tracers, some middle management and other employees, they also generate a demand for new skills, which are everywhere in short supply, and for new types of information within firms and from specialist firms in the provision of business and computer services (see especially volumes 3 and 5 of the Gower Series).

(7) The capability which IT confers to improve the quality of products, processes, and services. On-line monitoring and control

of quantity and quality of output has already led to dramatic improvements in industries as diverse as colour television and passenger cars. It leads to capital-saving, labour-saving, materials-saving and energy-saving improvements in production processes since it reduces the number of rejects and wasted components both for intermediate and final output. This reduces the requirements for inspection and lower management employees, but increases the requirement for skilled systems designers and engineers and the level of responsibility and skills for maintenance and for some types of operative (see volume 4 of the Gower Series).

(8) The capability which IT confers to link up networks of component and material suppliers with assembly-type firms (as in the automobile industry) or with service firms (as in the hotel and catering industries). An equally important function which IT can perform is with respect to the linkages between producers, wholesalers and retailers as in the clothing industry. In both cases it is the combination and convergence of communications with computer technology (figure 4.2) which permits savings in inventories at all levels in the system, and especially in work-in-progress, and a far more rapid and sensitive response to (even daily) changes in consumer demand. (See volume 1 of the Gower Series for the clothing industry (Soete and Rush); volume 4 for the car industry (Jones, 1985)).

(9) The increased integration of manufacturing and service activities means that it does not make any sense to speak of a 'service' economy or a 'manufacturing' economy, since they are inter-dependent. This is evident, for example, in the tendency for suppliers of hardware or components to offer a 'service' or at least a subsystem incorporating many service elements. The tendency is clear on a small scale in the mushroom growth of 'instant print shops', which are partly in retail distribution and partly in the manufacturing (printing) industry. On a much larger scale it is evident in the supply of design, engineering and other information services, together with entire new plants. In the automobile industry there is likely to be increasing integration between distributor networks and manufacturing operations. In the service industries the growth of software design capability is leading these industries into the same category of R and D performers and sources of innovation as the manufacturing R-and-D-intensive sectors have been for a long time. In chapter 10 we shall make use of the concept of an 'IT sector' comprising both IT-producer and IT-user sectors. In each case these include both manufacturing and services.

(10) Greater international integration of industries, services and markets as a result of much more rapid transmission of information

and vastly improved communication flows. This has already had big effects in the communication between offices, production sites and laboratories of large multinational enterprises. Accelerated international transfer of technology and greater mobility of service industries are among the probable consequences of these developments. The financial services industry is already demonstrating the far-reaching nature of these changes, and many countries and regions will ultimately be affected by major changes in the international division of labour and comparative advantage with the new technologies (see volume 5 of the Gower Series). Chemical engineering contractors are another industry to make external use of IT to smooth the workload between their various international sites and integrate design and procurement on an international basis.

The listing of these characteristics of the new paradigm is not intended to be exhaustive, but to illustrate some of the main ways in which it is affecting the entire economic system and not just particular products or sectors. It is for this combination of reasons that the shift to information technology involves major problems of structural change.

By contrast, the newer biotechnologies, although they certainly also have enormous future growth potential, have not yet reached the point where their macroeconomic impact could be great enough to carry the entire economy forward in the next decade or two. This illustrates the importance of the Mensch (1975) debate. It is the diffusion of the innovations of the fifties, sixties and seventies rather than the first innovations of the eighties which must provide the main impetus for a new economic upswing in the next two decades. The new biotechnologies can provide very important auxiliary growth areas and they may ultimately revolutionize agriculture, the food industry and the chemical industry; but the main elements of the new techno-economic paradigm do not come from this source.

Still less can nuclear technology play this role. Its applications are extremely limited and its capital cost is astronomical, so that any large programme would severely aggravate any capital shortage problems. Its cost advantages are even now dubious and there are strong environmental, social and political arguments for limiting its diffusion, especially in the case of the fast-breeder reactor (Surrey and Thomas, 1982; Walker, 1984). In addition, as a result of a wave of energy-saving innovations in the seventies and eighties the rate of growth of energy consumption has fallen. Although there is certainly a case for continuing with long-term research programmes in relation to solar power, wind power and fusion, it is unlikely that new cheap energy sources will provide a major new impetus to economic growth in the near future, especially in view of the fall in oil prices in 1986. Much more important are likely to be energy-saving technologies associated with microelectronics based monitoring, metering and control systems (Walker, 1984).

The Crisis of the Old Paradigm

Whilst any one of the sector studies which we undertook would illustrate the point about the emergence of a new techno-economic paradigm, there are of course innumerable other types of technical change, which affect particular processes and products, many of them important for individual industries. But for the period from the fifties to the seventies, many of these could be classified as improvements in mass and flow production techniques to produce very large quantities of standardized, homogeneous commodities, such as vehicles, consumer durables, steel and synthetics. At the same time, though, combined social, technical and organizational advances permitted the exploitation of the economies of scale associated with this older energy-intensive paradigm in many different industries and services. This path of development yielded diminishing returns in the sixties and seventies and as we will show in detail in chapters 7 and 8, capital productivity fell in many cases.

Although our sector studies have identified a large number of innovations associated with the computerization of production systems, offices and services, they also indicate a relatively slow rate of diffusion of these innovations outside a few leading edge industries. Moreover the pattern of diffusion is often of a type which cannot yield the full productivity gains which are potentially available from the new technology. For example, often only one new piece of equipment is installed in an office or machine shop which is otherwise still geared to an older technological style, or the maintenance system for the new equipment may not be capable of coping with the new technology, or the management may not understand how to exploit the technology for profitable production. Even when the intention of this new investment is a laudable attempt to use the most advanced technology available, this fragmented approach may actually lead to a fall in both capital and labour productivity. The rather fundamental nature of the paradigm change means that there are many problems of mismatch, both in the capital stock and in the labour force.

Only where the necessary skills have been assembled and when thorough training has been carried out at all levels, and when a systems approach to the entire design and production system is adopted and much new investment has been undertaken can the full potential benefits be realized. The diffusion in depth has hardly begun in many sectors, and there are many institutional changes which are needed before the full productivity and profit potential of the new technologies can be realized. What is at issue is a paradigm change and not simply incremental change. (See for example the discussion on diffusion of CNC machine tools, CAD and robotics in volume 4 of the Gower Series, or of the computerization of banking and insurance in volume 5.) This means that for a considerable period there is a 'mismatch' problem with respect both to skills and management styles and with respect to the capital stock. These mismatch problems give rise to major problems of structural unemployment and adjustment. A dramatic example is that of the UK newspaper industry. In part III and particularly in chapter 11 we discuss in greater depth the empirical evidence on paradigm change and in part IV the policies needed to facilitate this change. In chapter 10 we show that in some leading IT-user industries, as well as the IT-producer industries, considerable changes have already been made, but the diffusion process still has a long way to go.

II

METHODOLOGY AND

MODEL SIMULATION

The question of how to model the process of technical change raises some fundamental challenges, both from the theoretical model-building and from the more applied 'forecasting' perspective. As illustrated in the previous chapters, the processes determining the rate and direction of technical change are complex and poorly understood. They are based on factors transcending the purely economic field and including predominantly technological, institutional or even 'organizational' factors. As a result, few attempts have been made to incorporate technological change in a direct way in economic models; preference seems to go for a simple 'time related' representation, often based on extrapolations of past productivity growth.

Such treatment is particularly at loggerheads with our own more 'Schumpeterian' approach to the process of technological change, in which 'clusters' of inter-related innovations, 'band-wagon' investment booms and 'creative capital' destruction are the main features and the prime causes of dynamism and structural change in the economy. Unfortunately the modelling challenges raised by such an approach are even more formidable. Available technological proxies, such as those used in the chapters in the ensuing empirical part of this volume, will be of little direct use in modelling any of these effects. With respect to the employment impact of technical change, the relevant question is indeed not one of modelling 'advances in technology', but rather the application; the extent to which these advances have been translated in achieved productivity levels. Without information on the speed by which new technological advances are being adopted, there seems little scope for any formal incorporation of technological change (whatever proxy one may wish to choose) in dynamic economic models.

One approach which has tried to model some of the 'adoption' aspects of technological change is the vintage-capital approach. Whereas the basic assumption behind such an approach is that actual diffusion is assumed to be instant across all new investment,[1] it nevertheless offers a modelling framework which is nearest to our understanding of the process of technological change. A fundamental notion behind this approach is that a very important component of technical change is that associated with, or 'embodied' in, plant and equipment. Technical progress is linked directly with the process of capital accumulation; each successive investment decision may involve the purchase of equipment of a type qualitatively different (presumably more productive) than those already in use, and hence any gross investment (even if purely to replace existing equipment) tends to raise productivity. The emphasis is on the factors determining the speed at which new techniques are absorbed into the production system, on the importance of the investment process in raising productivity, and on the fact that overall efficiency depends on the history of technical change, investment scrapping, etc.

We start this methodological part with a discussion of some of the theoretical problems involved in incorporating technical change in a traditional investment function framework. Particular attention is paid to the Cambridge Growth Project model, the only large-scale UK model, in which productivity increases are introduced via new investment to forecast (un)employment. In chapter 6 an alternative model is put forward, in which the technological parameters, rather than being assumed a priori, are the actual outcome of the model. Finally, in chapter 7 and as a precursor to the subsequent empirical part, we present some first aggregate model simulation results for the UK manufacturing sector. The manufacturing sector lends itself well to such a simulation exercise, present levels of capital accumulation embodying 'best-practice' technology providing probably the best 'limit' as to future productivity and employment growth. Readers who do not wish to examine these methodological problems should proceed directly to part III.

Note

1 As Hahn and Matthews (1967, p. 60) put it: 'the manna of technical progress falls only on the latest machines'.

5
Forecasting Investment in the Medium Term

Robert Eastwood

Economists have always recognized that investment and innovation were interconnected and essential features of the process of economic growth.[1] The introduction of new products requires new investment, often, but not always, located in new factories and plants. The improvement of established processes of production, through various forms of automation or new instrumentation, also usually requires investment in new types of capital goods. Some technical change occurs continuously without new investment, simply through an improved understanding and more efficient utilization of existing plant and machinery. This is sometimes referred to as 'disembodied technical change'. But it is nevertheless a reasonable generalization that technical change in modern industrial societies depends upon a continuous flow of new investment, deepening and extending the capital stock of the economy. This is especially true of the manufacturing sector, but as will become clear in chapter 9, it is also true of major capital-intensive non-manufacturing industries, such as energy, communications and transport.

Common sense suggests that technological change should affect investment in several ways. Most obviously it should make prospective investments appear more profitable and thus have an expansionary effect. On the other hand, if it is capital productivity which is raised, then the investment required for a given increment to output will be less. Most studies of investment appear, though, to give little weight to technological change variables.

This chapter is concerned with the medium-term modelling of investment, particular attention being paid to the role of technological change. The focus is on the Cambridge Growth Project model, the only disaggregated input–output model for the United Kingdom, and as a consequence, the cornerstone of quite a number of technology–employment analyses and

73

forecasts. The discussion here is limited to the underlying theoretical framework used in the investment and employment blocks of the model. The plausibility and actual 'performance' of the model in forecasting employment is picked up in chapter 11.

We start this chapter by setting out the basic theoretical treatment of investment demand in a non-vintage framework. Certain issues which arise in both vintage and non-vintage models can be discussed most clearly in the simpler framework. In the next section, we then present a critical account of the modelling of investment and technological change in the Cambridge Growth Project model. It is shown that – at least in its 1976 incarnation – theoretical errors in the model probably lend a pessimistic bias to the forecast consequences of technological advances.

Some Theoretical and Methodological Preliminaries

A Basic Investment Equation

In this subsection an investment function is derived under very simple assumptions and the legitimacy of collapsing this into the standard accelerator model is examined. Brief discussion is given to dropping assumptions other than those which will be dropped in the rest of the paper.

The following notation will be used throughout this chapter.[2] $K(t)$, capital stock at t; $L(t)$, labour input at t; $I(t)$, gross investment at t; $p(t)$, output price at t; $q(t)$, price of machines at t; $w(t)$, wage rate at t; $r(t)$, interest rate at t; δ, depreciation rate of capital; $F(K(t)\exp(\lambda^1 t), L(t)\exp(\lambda^2 t))$, production function; $Z(p(t))\beta(t)$, output demand function; ϵ, price elasticity of output demand, defined positive; $M_t(p(t))$, marginal revenue; $c(t)$, cost of capital (dependence on other variables depends on context); $Y(t)$, output; λ_1,λ_2, rates of (disembodied) capital and labour augmenting technical change.

δ is only used where constant depreciation is being assumed. The output demand function facing the firm is always supposed to have the separable form shown; ϵ, the price elasticity of output demand is to be interpreted as the elasticity facing (or perceived by) the individual firm; $\beta(t)$ is a shift variable representing demand.

Listed below are seven assumptions which allow the derivation of a simple investment demand equation:

  A1, perfect capital market;
  A2, no uncertainty;
  A3, constant returns to scale in production; no inputs other than labour and capital;

A4, the characteristics of machines are independent of their age and birthdate, although they do depend on time, and machines evaporate at rate δ;

A5, competitive factor markets;

A6, no taxes of any description;

A7, profit maximization.

Assumption A4 means that we can refer unambiguously to 'the' stock of machines at any date, and can write:

$$\dot{K}(t) = I(t) - \delta.K(t). \tag{5.1}$$

Assumption A5 implies that the firm can buy or sell machines at any date instantaneously and without incurring transactions costs. Moreover the price of machines is fixed and known in advance (A2). This makes the investment decision of the firm pseudo-static inasmuch as the amount of capital that the firm both desires to hold and will hold at time t is independent both of its capital inheritance, $K(t-1)$, and of all future prices other than $q(t+1)$. Indeed, its capital and labour use, at date t, can be derived from the following 'marginal revenue product' conditions:

$$M_t(p(t)) \frac{\delta F[K(t)\exp(\lambda_1 t), L(t)\exp(\lambda_2 t)]}{\delta L(t)} = w(t) \tag{5.2}$$

and

$$M_t(p(t)) \frac{\delta F[K(t)\exp(\lambda_1 t), L(t)\exp(\lambda_2 t)]}{\delta K(t)}$$

$$-q(t)\left[r(t) + \delta - \frac{\dot{q}(t)}{q(t)} \right] \tag{5.3}$$

where we can denote the right-hand side (RHS) of equation (5.3) simply by $c(t)$. Thus for given paths for $q(t)$, $w(t)$, $r(t)$ and $\beta(t)$, the whole production programme for the firm through time can be derived. One may think of the firm as first choosing the capital:labour ratio that minimizes unit costs at each date (formally by eliminating $M_t(p(t))$ by dividing equation ((5.2) into (5.3)) and then choosing a scale of operation that equates unit costs to marginal revenue. Output price and quantity along with inputs are thus jointly determined at each date.

Implicit in the above is gross investment at each date, and the following expression can be derived:

$$\frac{I(t)}{K(t)} = \frac{\dot{\beta}}{\beta} + \delta - \lambda_1 + (1 - \epsilon_K)(\sigma - \epsilon) \left(\frac{\dot{w}(t)}{w(t)} - \lambda_2 \right)$$

$$- [(1 - \epsilon_K)\sigma + \epsilon\epsilon_K] \left(\frac{\dot{c}(t)}{c(t)} - \lambda_1 \right). \tag{5.4}[3]$$

Here σ is the elasticity of substitution between effective capital and effective labour in the production function. ϵ_K is the elasticity of output per head with respect to effective capital per head.

One can see from equation (5.4) that if factor prices change by just enough to offset technological advances, and if there is no change in demand, effective input levels will be held constant, and investment will be just sufficient to make up for depreciation net of capital-augmenting technological change. If, however, factor prices change, then investment will be affected in two ways.[4] First, a relative change will induce the firm to change the technique of production so that investment will be changed even if output is not. Second, an absolute change resulting in a fall in the minimized cost of production will cause a fall in output price and an expansion down the demand curve for output. It is, in fact, possible to rewrite equation (5.4) to reveal the resolution of investment into components reflecting replacement, substitution and scale effects:

$$\frac{I(t)}{K(t)} = (\sigma - \lambda_1) + \sigma(1 - \epsilon_K) \left[\left(\frac{\dot{w}(t)}{w(t)} - \lambda_2 \right) \right.$$

$$\left. - \left(\frac{\dot{c}(t)}{c(t)} - \lambda_1 \right) \right] + \frac{\dot{Y}(t)}{Y(t)} \tag{5.5}$$

equation (5.5) becomes a simple accelerator process, then, if relative (effective) factor prices are unchanged.

The Assumptions Underlying the Basic Investment Equation

Equations (5.4) and (5.5) are obviously heavily dependent on some quite unacceptable assumptions among A1–A7, as was evident from chapter 3. The objectionable second part of A3 is easily circumvented, and much of the rest of this chapter is concerned with dropping A4 and an unacceptable component of A5. Beyond these, empirical work has to pay serious attention to A1, A2 and A6. In this chapter, however, the problems raised by these three assumptions are not discussed at any length. These are, in a sense, peripheral and self-contained topics, excellent discussions of which are to be found in Nickell (1978).

Much of the rest of this chapter will be concerned with the consequences of modifying A4. Even to relax the depreciation assumption alone means that information about the firm's capital can no longer be adequately captured by an aggregate notion, since the aggregate rate of depreciation will depend on the age structure of the firm's capital. Since the evidence suggests neither that exponential decay is a good assumption nor that the age structure of the capital stock is anything like constant, a serious problem is created which seems to have been ignored in econometric studies.[5] Turning to technical change, it may be noted here that, once the idea of embodiment is introduced, a pseudo-static formulation of the investment function becomes impossible because in evaluating the acquisition of a unit of capital at t, the firm is obliged to calculate the present value of the stream of quasi-rents that it will yield, as opposed to the return just in the current year. This supposes, reasonably, that second-hand capital goods are not freely marketable. A further complication is that the investment equation now has to take into account not only depreciation of inherited capital equipment, but obsolescence of old vintages.

The critical component of assumption A5 is the stipulation that there be a perfect market in capital goods. It is this assumption, indeed, which allows a pseudo-static formulation. The mildest way of relaxing this is to make investment decisions irreversible, for this would only become relevant if circumstances were such that the firm actually wished to disinvest. Beyond this one may allow either for costs of, or lags in, adjustment. Except in special cases, either generalization results in a dynamic investment equation in which distributed lag functions of investment and/or output appear on the right-hand side. The possibilities here are numerous and are well surveyed by Nickell (1978, chapters 3, 4 and 8.4(ii)).

Investment in the Cambridge, and Other, Vintage Models

In discussing the work of the Cambridge Growth Project on investment, there are three difficulties. First, the Cambridge model has gone through at least four versions and this quintessentially public good was an early casualty of privatization. The result is that the analysis here has mainly to be based on the 1976 version (Barker, 1976) of the model, although there is some evidence in Barker et al. (1980) that by 1980 little had changed in so far as investment was concerned. Second, although the investment function (in Barker, 1976) purports to have a theoretical foundation, its derivation is rather informal and there are various inconsistencies among the assumptions. Third, although investment by the firm must be considered as jointly determined with its labour input, output and (if the product market is not competitive) output price, we find that the assumptions used in the employment chapter are inconsistent with those in the investment chapter. The best way to proceed, therefore, seems to be to provide first an introduction to the vintage model, and subsequently to consider the Cambridge work.

Investment in the Vintage Model

Vintage models arise when we relax assumption A4 to allow the characteristics of machines to depend not only on calendar time, but also on their date of installation. It is also assumed that there is no question of selling a machine once installed. The technology may be assumed to be putty–putty, putty–clay or clay–clay according to whether substitution between capital and labour is permitted both *ex ante* and *ex post*, just *ex ante*, or not at all. The exposition here assumes the putty–clay technology.

In the absence of adjustment costs or lags, optimum behaviour by the firm can be described in terms of three conditions involving first, the amount of capital inherited from the previous period that is still worth using, second, the factor proportions to be used on new equipment and third, the scale of new investment. Thus:

1 Scrapping: scrap (or lay off) any capital that cannot earn a non-negative quasi-rent.[6]
2 Factor choice: select new vintage capital intensity to minimize a suitably constructed discounted sum of unit costs.
3 Scale: ensure that investment is on a scale such that the present value of investment in one more machine would be zero.

No investment equations that embody condition (1) exist. The standard practice is to lump obsolescence in with depreciation so that productive capacity decays at rate δ in the absence of new investment. Implementation of condition (2) depends on whether or not there is embodied technical change. If there is not, then the prospective economic life of a new machine is infinite.[7] The total cost, $x(t)$, to be minimized, of providing a decaying infinite stream of output – one unit at date t and $\exp(-\delta(s-t))$ at dates s thereafter – is:

$$x(t) = q(t) \, v(t) +$$

$$\int_{s=t}^{s=\infty} \exp(-(r+\delta)(s-t)) \, w_e(s,t) \, 1\,(s,t)\,ds \qquad (5.6)$$

or

$$x(t) = q(t)\, v(t) + \overline{w}(t) 1(t) \tag{5.7}$$

where:

> $1(t) =$ labour per unit output on a newly installed machine at date t,

> $v(t) =$ capital per unit output, similarly,

> $1(s,t) =$ labour per unit output on such a machine at date s,

> $w_e(s,t) =$ expectations at date t of wages at date s,

> $\overline{w}(t) =$ a generalized expected wage, defined so that the RHS of equation (5.6) equals the RHS of equation (5.7).

If there is embodied technological change, the upper limit of the integral in equation (5.6) has to be replaced with the date of expected obsolescence, and the implied economic life of the machine must be determined jointly with capital intensity (Ando et al. 1974). Either way, given the *ex ante* production function at date t, minimization of equation (5.7) results in an optimal capital:output ratio on new equipment $v^*(t)$, which depends on $(q(t)/\overline{w}(t))$, production function and technical change parameters – which can be lumped together as a parameter vector Θ – and time. Thus we may write:

$$v^*(t) = v^* \left[\frac{q(t)}{\overline{w}(t)}, \Theta, t \right] \tag{5.8}$$

$1^*(t)$ is determined similarly as a function of the same variables.

Associated with this optimum is a minimized discounted expected unit cost, $x^*(t)$, and so condition (3) gives us:

$$x^*(t) = \int_{\tilde{s}=t}^{s=\infty} M(p_e(s,t)) \exp(-(r+\sigma)(s-t)) ds \tag{5.9}$$

where $p_e(s,t)$ is defined in the same way as $w_e(s,t)$.

Assuming a demand elasticity that is constant through time, this becomes:

$$x^*(t) = \int_{\tilde{s}=t}^{s=\infty} (1-\tfrac{1}{\epsilon}) p_e(s,t) \exp(-(r+\sigma)(s-t)) ds. \tag{5.10}$$

Equation (5.10) implicitly allows us to solve for $p(t)$ – supposing that $p(t)$ equals $p_e(t,t)$ – and provided that we specify price expectations. So one may write:

$$p(t) = p(q(t), \overline{w}(t), \Theta, \epsilon, t). \tag{5.11}$$

This equation is obtained, in principle, by substituting equation (5.7) – with x, v and 1 taking their starred values – into equation (5.10) and rearranging. Any dependence of expected prices on variables other than those appearing in equation (5.11), and any parameters of the expectations mechanism, are subsumed in the functional form of that equation.

We now have, in principle, factor choice and output price determined. Then, similar to the discussion following equation (5.3), output at t is obtained from the demand function:

$$Y(t) = \beta(t)\, Z(p(t)). \tag{5.12}$$

Denoting the optimal output by $Y^*(t)$, we finally obtain:

$$1(t) = v^*(t) \left[Y^*(t) - (1-\delta)\, Y(t-1) \right]. \tag{5.13}$$

Having established the general structure of the vintage model, we can now turn to its elaboration in the Cambridge Growth Project.

Investment and Employment in the Cambridge Growth Model

In chapter 5 of Barker (1976), a number of assumptions are made which are not mutually consistent:

B1, each industry is viewed as a single 'firm' which minimizes costs subject to an output constraint;

B2, the technology is putty–clay; there is no depreciation of equipment so that a machine once installed has capital and labour productivities that are fixed for ever;

B3, machines are scrapped when their quasi-rents fall to zero;

B4, machines are expected to be scrapped after a fixed time T^*. The capital intensity of new equipment is chosen so as to minimize the expected cost of operation over a horizon T^*;

B5, a fraction δ of each year's potential output is lost at the end of the year as a result of scrapping.

Taking B1 and B2 as primary assumptions, it is clear that B3 and B5 are generally inconsistent with them and with each other. B3, for instance, is a consequence of profit maximization, but not of cost minimization. B4 is not inconsistent if one is allowed to 'rig' wage and price expectations freely, but otherwise is, in general.

Selecting from these assumptions as is convenient the analysis proceeds as follows. B5 and B1 tell us the amount of output, $\overline{Y}(t)$ that has to be produced by the new vintage at date t. Knowing this, and given a CES production function with returns to scale parameter k:

$$\overline{Y}(t) = A \exp gt \left[f\overline{L}(t)^{-\varrho} + (1-f) \hat{I}(t)^{-\varrho} \right]^{-k/\varrho} \qquad (5.14)$$

where there is Hicks-neutral technological progress at rate g, $\overline{L}(t)$ is employment on the new vintage, $\hat{I}(t)$ is 'desired' investment, $p = \frac{1}{1}$ where σ is the elasticity of substitution, and f is the share parameter, the cost-minimizing conditions can be written:

$$\lambda(t) \ \frac{\partial \overline{Y}(t)}{\partial \hat{I}(t)} = q(t) \qquad (5.15)[8]$$

$$\lambda(t) \ \frac{\partial \overline{Y}(t)}{\partial \overline{L}(t)} = \overline{w}(t) \qquad (5.16)$$

where $\lambda(t)$ is the shadow price of output.[9]

Equations (5.14), (5.15) and (5.16) correspond to 5.11, 5.8 and 5.9 on p. 113 of Barker (1976). They also correspond to the conditions necessary for the minimization of equation (5.7) above, so the correct way to proceed is to eliminate $\lambda(t)$ between equations (5.15) and (5.16) and use the derivatives of equation (5.14) to obtain the cost minimizing values for $(\hat{I}(t), \overline{L}(t))$ as functions of $q(t)$, $\overline{w}(t)$ and $\overline{Y}(t)$. Note here that since constant returns to scale are not assumed, one cannot generally obtain optimal input coefficients as functions of input prices independently of scale. What in fact is done is that equation (5.16) is forgotten about and for $\lambda(t)$ is substituted a discounted sum of expected output prices over the horizon T^*:

$$\lambda(t) = \int_{s=t}^{s=t+T^*} \exp(-r(s-t)) \ p_e(s,t) ds \qquad (5.17)[10]$$

Given this, an investment equation can be derived from (5.15):

$$\ln I(t) = a_0 + a_1 t + a_2 \ln \frac{q(t)}{\lambda(t)} + a_3 \ln \overline{Y}(t). \qquad (5.18)[11]$$

It is then assumed that $p_e(s,t) = p(t)$, all s, so that $\lambda(t)$ can be replaced by $p(t)$ with a term involving r and T^* disappearing into the constant a_0. Adding an adjustment lag, one obtains:

$$I(t) = \gamma \hat{I}(t) + (1 - \gamma) I(t-1). \tag{5.19}$$

It is then possible to arrive at the final estimating equation:

$$I(t) = (1 - \gamma)I(t-1) + \gamma \exp(a_0 + a_1 t) \left[\frac{q(t)}{p(t)} \right]^{a_2}$$

$$\left[Y(t) - (1 - \delta) Y(t-1) \right]^{a_3}. \tag{5.20}$$

This equation is estimated for investment by each sector in each asset and it would in fact be possible to extract estimates for g, f, p and k from the a_1, although this is not done. Inspection of a selection of the results (Barker, 1976, pp. 116–17) shows that while most coefficients are correctly signed, many are insignificantly different from zero. Estimates vary very widely and implausibly. Much of the variation in $I(t)$ is 'explained' by $I(t-1)$. This latter feature creates a difficulty in forecasting since there is a danger that the starting value of I and the short-term dynamics will dominate the medium-term forecast. This is avoided by solving out a 'steady-state' investment equation from equation (5.20). This is essentially, when linearized, a simple accelerator with a cost of capital term and a time trend thrown in.

When looking at the employment chapter of Barker (1976), one finds a treatment of production which is not consistent with the model just outlined. Before discussing this, however, it seems appropriate to make a number of comments on the present model. First there is the question of demand and prices. Whereas the theoretical analysis given earlier in the paper shows that the derivation of a price equation under profit maximization is an essential intermediate step in obtaining an investment equation, we do not find this in the Cambridge model. Prices are in fact modelled quite separately (Barker, 1976, ch. 9) as a mark-up on industry average variable costs and for forecasting purposes the outputs $Y(t)$ and $Y(t-1)$ can then be obtained from the demand side of the model. To combine this procedure with the assumption of cost minimization is perhaps consistent,[12] but what is not is to then use equation (5.17) to obtain $\lambda(t)$. This would only be correct if prices (and wages) happened to be such that an industry of competitive producers would in fact have arrived at output $Y(t)$ and price $p(t)$ on the basis of profit maximizing behaviour. We have seen earlier that, if the elasticity of demand is constant, price should be some kind of mark-up on minimized unit cost on new equipment (see equation (5.13)). If this is replaced with a constant mark-up on industry average cost, simulations in which anything of interest changes on the production side (such as the rate of technological change) will provide misleading results. The claim in Barker (1976, ch. 5) that producers 'do not expect to make windfall profits as a result of new investment'[13] can, as a matter of fact, not be sustained in view of the actual way in which prices are modelled.

In order to emphasize the potential importance of the point made and the nature of the biases involved, one can sketch out how the Cambridge model would react (in forecasting mode) to an increased rate of labour-saving technical change, assuming no changes in wages. The direct effect would be a small fall in prices, variable costs having only fallen on new equipment. This would have two offsetting first round effects on investment. The substitution effect would tend to lower investment (not if $q(t)$ fell by as much as $p(t)$), while the demand effect would tend to raise it. One may surmise that the net effect on output would be small, whichever way it went. There would probably, however, be a fall in employment. In other words, the Cambridge model, by ruling out a boom led by investment in highly profitable new equipment and sustained by movements down the demand curve for output, is bound to present a pessimistic assessment of the consequences of faster technical change. Altering the model to allow effects of this kind, however, would be a far from trivial matter. It is not simply a question of leaving the price equation as it is and correcting the investment equation to allow for the profitability of new investment. If the output produced with the new capital is to be sold, prices must be allowed to fall, and so the assumption of a constant mark-up on industry average variable costs has to be dropped. Apart from raising theoretical questions,[14] modifications along these lines might require changes in the solution algorithm of the model.

There are other limitations of the treatment of investment. First, there is no depreciation. Obsolescence appears in the form of a constant decay factor δ on potential output. Second, price expectations are static and over a fixed horizon T^*. Third, only embodied Hicks-neutral technical change is permitted. Fourth, while variable returns to scale are permitted the empirical outcome is unreported. We now turn to the treatment of technical change in the employment 'block' of the Cambridge Growth Model.

In chapter 8 of Barker (1976) the link from investment to employment is made in a very simple way.[15] It is assumed that the economy has for ever been enjoying steady state growth with Harrod-neutral technical change and, as a consequence, constant factor shares. From the data, trend rates of growth of output and labour input, and the share of wages, can be obtained. Given the scrapping criterion B3, the economic life of equipment, T^*, can immediately be obtained as can the capital:output ratio on equipment (cumulative investment over the past T^* years divided by current output).[16] The average labour:output ratio allows one to infer labour:output ratios on all vintages in use. Denoting 'scrapped' output as t by $S(t)$, we may see how the employment forecast is obtained as follows. We have:

$$\overline{Y}(t) = I(t)/v(t). \tag{5.21}$$

By definition

$$S(t) = \overline{Y}(t) - \left[Y(t) - Y(t-1) \right] \tag{5.22}$$

$$L(t) - L(t-1) = l(t)\,\overline{Y}(t) - l(t - T^*)S(t). \tag{5.23}$$

Using equation (5.21) in (5.22) and then equations (5.21) and (5.22) in (5.23), we obtain the change in employment as a function of the known capital:output and labour:output ratios, $I(t)$, $Y(t)$ and $Y(t-1)$.[17,18] In relation to our focus on investment two points need to be made.

First, the lack of integration between the treatment of output and investment on the one hand and employment on the other is noteworthy. Technical change is modelled differently, being Harrod-neutral in the employment chapter and Hicks-neutral in the investment chapter. It may be observed that had Harrod-neutrality been assumed in the investment chapter, the time trend would have disappeared from the investment equation (damaging without doubt the goodness of fit), while Harrod-neutrality was necessary in the employment chapter for the steady-state assumption to be permissible.[19]

Second, there is a significant lack of statistical testing in the employment chapter. The Cambridge researchers implicitly take *here* the position that no information of interest to medium-term forecasting can be derived from econometric work on short-term data.[20] One is therefore left only with time trends for output and labour input and the stylized fact of a constant income distribution. With only three pieces of information one can evidently infer little and must in consequence assume much. There is certainly no room for degrees of freedom.

As remarked earlier on, the Cambridge model has gone through some modifications since 1976 and the fourth version (MDM4) is described in another paper by Barker et al. (1980). The crucial change is that the model is now dynamic and generates simulation paths for the variables of interest rather than merely forecasts at a target date. Although the 1980 paper presents a 'condensed form' of the model, only simulations from the full version are reported. It would appear from these simulations and accompanying remarks, that little has changed as far as investment is concerned. In particular it is shown quite explicitly that while a faster growth in world trade can cause an investment boom, changes in 'time trends' and the rate of change of labour productivity cannot. One may surmise that the mechanisms by which such an effect might be produced remain arbitrarily blocked.

Notes

1 This chapter is an abridged version of a larger paper with the same title, obtainable from the author on request.
2 Notation will be introduced as needed. When time arguments are dropped, it is to be inferred that the variable is assumed constant.
3 cf Nickell (1978), p. 18. My equation simply generalizes his 2.19 to allow for technical change. The derivation is analogous to that provided by him on p. 22.
4 The word 'effective' is omitted from the next few lines for stylistic reasons.
5 For a summary of evidence on depreciation and the age structure of capital, see Nickell (1978), pp. 304–8.
6 Defined as marginal revenue minus unit operating cost.
7 I assume here that the firm plans positive gross investment at every future date.
8 I have left out fiscal adjustments involving the cost of capital.
9 (5.14), (5.15) and (5.16) correspond to 5.11, 5.8 and 5.9 of Barker (1976).
10 Equivalent to (5.10) in Barker (1976).
11 Equivalent to (5.12) in Barker (1976).
12 If hardly very appealing.
13 Barker (1976), p. 113.
14 The assumption of a constant rate of depreciation/obsolescence would become difficult to accept in these circumstances.
15 For a full description of the procedure, see Wigley (1970).
16 Remembering that there is no depreciation.
17 cf equation 8.8 on p. 192 of Barker (1976).
18 We may note that this procedure has certain disadvantages owing to its reliance on steady-state assumptions.
19 The estimates in the two chapters have implications for one another which are not explored. Thus estimates of the obsolescence rate and the equivalent lifetime could be unscrambled from each set of estimates and compared (a value being assumed for the discount rate). This is not done.
20 While in the investment chapter they *are* willing to base medium term investment forecasts on an investment function fitted to short-term data.

6

A Vintage–Capital Simulation Model

John Clark

Having offered a detailed critique of the underlying Cambridge Growth Model 'vintage' framework, we set out in this chapter an alternative vintage 'simulation' model. As indicated in the introduction to this section, we regard this kind of model as particularly appropriate to analysis of technical change. Acknowledging the difficulties outlined in the previous chapter, however, we adopt a relatively simple and pragmatic approach. The principal objectives of the model described below can be summarized as follows:

(1) The first objective is to try and use historical economic data to describe how rates of technical advance have changed over time and become embodied into the production system. These changes are then interpreted with reference to known innovations and technological diffusion. To this end the concept of 'best-practice productivity' or 'marginal productivity' is used to describe quantitatively the development of the technological frontier over time, a concept closely allied to the potential available from innovations. The realization of this potential depends on the level of investment in the new technologies, whereby overall efficiency of production may increase rapidly for many years following a radical innovation even if no further important innovations are forthcoming.

(2) The second objective is to use the derived spectrum of technologies in use, as embodied in fixed capital, together with estimates of future rates of technical change, to relate output growth to employment; for given levels of demand, the future of employment depends on the heritage of technologies in use as well as on new technologies, investment, and utilization rates. A feature of the model is the avoidance of any assumption of constancy in historical rates of technical change.

86

In addition, quantities such as utilization rates, the age of the capital stock and 'underlying' (trend) rates of productivity growth are estimated. With its specifically supply-side orientation, the approach here is intended to be complementary to demand-orientated macroeconomic models.

This work fits into the class of so-called 'vintage' models of sectors of manufacturing industry, which have been fairly prevalent in the economic literature over the last decade or so. The objectives and approach adopted here differ, however, from those of most of the work previously published; it is less formal than econometric exercises designed to provide a 'best fit' to observed historical growth patterns, the latter generally requiring the use of assumptions, such as constant rate of technical change, which our analysis seeks particularly to avoid. The general approach described here is therefore closer to those such as Nelson and Winter (1982) in their simulation studies of industrial evolution, or Soete and Turner (1984) in their macroeconomic diffusion model.

We therefore prefer to adopt an approach in which actual data (on, e.g., output and employment) are used as inputs, and the validity of the technological outputs obtained is assessed by comparison with other studies and by discussion with individual industrial experts. The analysis is based on the notion of 'vintages' of capital equipment; it is assumed that capital goods purchased in different years in a particular industrial sector have different technological characteristics, and hence reflect, or 'embody', improvements in technology over time. In particular, the labour and capital required to produce one unit of output is allowed to vary between equipment installed in different years, so that the capital in any industry has a structure, with more recent equipment, in general, having a higher associated productivity than older equipment. Aggregate economic quantities such as the overall level of labour productivity can be obtained from weighted sums of the corresponding quantity taken over all vintages of capital equipment in use.

In this analysis it is assumed that the labour and capital inputs required per unit of production are fixed over the operating life of any particular vintage of equipment; once installed, there is no possibility of substitution between these factors of production. The possibility that, for example, an increase in wages relative to rental on equipment might induce businesses to attempt to reduce their use of manpower per unit of output at the expense of less productive use of machinery is therefore not incorporated. This simplifying assumption is not unusual and represents the view that the possibilities for such substitution are in fact quite limited (see chapter 3), which, as we have seen, is true of many IT technologies. With regard to the *ex ante* fixity of technical coefficients the current approach cannot strictly be called 'putty–clay' or 'clay–clay' since it attempts to derive an approximation to the actual technologies adopted in each year, not those expected given price movements and some selection rule (putty), nor those assumed a priori (clay).

The advantages of disaggregating the capital stock into vintages relate primarily to the fact that changes in overall industrial productivity can now be split into components; in any particular year, productivity may be enhanced by significant technological advances as reflected in large changes in the technological coefficients of the equipment coming onstream in that year, but the overall impact of this advance will depend also on the levels of investment undertaken in the more advanced equipment. As illustrated in chapter 7, this simple but frequently overlooked point may help to explain the apparent anomaly of low labour productivity growth in the seventies in many OECD countries (and in the United Kingdom in particular) despite the widely publicized availability of new technologies, in particular microelectronics, which are capable of bringing about dramatic productivity advances. Such technologies cannot be introduced overnight, and substantial initial investment outlays as well as much training may be needed to introduce them at all.

Changes in aggregate productivity can also come about from other factors not directly related to technology. The scrapping of equipment of less than average productivity (discussed below) will tend to raise the overall level, as will shifts in the composition of output whereby sectors of the economy with relatively high productivity expand more rapidly than low productivity sectors. Short-term 'labour hoarding' in periods of recession is accounted for in the model by variations in the 'utilization rate' of labour.

Description of the Model

As indicated, the model comprises a simulation of the temporal development of the output and employment potential associated with successive 'vintages' of fixed capital in a particular sector, or in manufacturing industry or the economy as a whole. An individual vintage or generation of equipment is that fixed capital installed during a particular year, and it is assumed that the technology 'embodied' in a vintage can be described by specific labour and capital productivities which remain associated with it throughout its working life. This is of course a major simplification, made to ensure tractability and interpretability of the model. Thus the potential output provided by a vintage is given by the level of investment in it, multiplied by its associated capital productivity; and the potential employment provided by it is given by this output multiplied by the corresponding labour:output ratio. Aggregate values of these variables are obtained by summing over all vintages in use.

In other words, if equipment installed in year τ has an associated output: capital ratio $b(\tau)$, then total output Q in year t can be written as:

$$Q(t) = U_K(t) \sum_{\tau=l}^{t} [\, I(\tau) - S(t,\tau)\,]\; b\,(\tau) \qquad (6.1)$$

where $I(\tau)$ is the magnitude of investment in year τ and $S(t,\tau)$ is the quantity of this investment scrapped by year $t(0 \leqslant S(t,\tau) \leqslant I(\tau))$. The expression under the summation sign gives the 'potential' output $Q^p(t)$ from surviving equipment; the capital utilization rate $U_K(t)$, assumed to have a common value for all vintages at time t, is then defined as the ratio of 'actual' to 'potential' output. In a similar way, a time-dependent variable $U_L(t)$, a measure of the utilization of labour, can be defined (see below).

It is assumed that data are available for the aggregates Q, L and I. Given also a time series for the capital stock, we have, together with an assumption on the scrapping pattern, e.g. that the oldest equipment is scrapped first, values for S. Some procedure is then required for deriving sets of technical coefficients and utilization rates such that these equations are satisfied. There is no unique way of carrying out such a procedure; any conclusions drawn from the model would ideally be robust for a wide spectrum of alternative 'choices' of utilization rates and technical coefficients, although in some cases we do have some direct information on the variables.

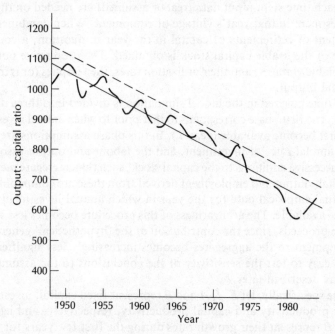

Figure 6.1 Average capital productivity.

The simplest approach is to assume that the output:labour and output: capital ratios $a(\tau)$ and $b(\tau)$ follow a linear trend reflecting the pattern of the observed aggregate labour and capital productivities, respectively. In the case of capital productivity, for example, figure 6.1 shows the observed trend in UK manufacturing industry, 1948–82. The straight solid line shows, for purposes of illustration, the overall trend in this ratio. If (as shown by the broken line) this is moved up until it becomes tangential to the observed capital productivity line, we may take as a measure of full utilization that level pertaining in the corresponding year. Annual estimates of the degree of under-utilization of capital can then be obtained by measuring the vertical distance between the observed trend and the broken line. Such an approach to estimating utilization rates has been adopted by Panic (1978).

This methodology is based on the assumption that, to a good approximation, the actual output–capital ratio would have followed a smooth linear path if fixed capital had been fully utilized throughout. This implies in turn that there are no changes in trend in the technical characteristics of new plant and equipment over time – in the notation used here, the $b(\tau)$ follow a smooth path reflecting the movement of the aggregate capital productivity. Here alternative, less rigid approaches will be investigated, in the hope that interesting changes in the rate and direction of embodied technology can be captured.

At each time step, input data (real or assumed) are needed on the level of investment in that year's vintage of equipment. When combined with the extent of retirements of capital in the year in question, a complete profile of the usable capital stock is obtained. The productive potential of available vintages, and their utilization rates, then implies (or is implied by) total output.

The time covered in the model simulation is divided into three distinct phases. The first phase corresponds to that prior to when annual investment data first become available (year T). In this phase assumptions are made about annual rates of investment, and the labour and output associated with successive additions to the capital stock, such that aggregate measures of capital, output and employment derived from these are compatible with the actual empirical date for the year in which annual investment series become available. The arbitrariness of this procedure becomes less serious as time proceeds, since the contribution of the 'hypothetical' generations of equipment to the aggregates becomes increasingly less significant; it is also easy to test the sensitivity of the conclusions to the assumptions used, as described later.

More specifically, let K, I, A and B represent fixed capital, investment, labour productivity and capital productivity, respectively, and let $\hat{K}$, $\hat{I}$, $\hat{A}$ and $\hat{B}$ represent their growth rates during the first few years for which annual series are available. During the first phase it is assumed that the capital stock and investment grow at rates $\hat{K}$ and $\hat{I}$ respectively; the notional scrapping rates during this phase can then be ascertained. It is then assumed that the marginal labour and capital productivities grow at rates $\hat{A}$ and $\hat{B}$ respectively. Marginal productivities are defined as those associated, at 'normal' utilization, with the hardware brought onstream during this phase. The absolute levels a and b of the technological coefficients are determined by ensuring that

$$Q(T) = \sum_{\tau=1}^{T} \left[I(\tau) - S(T,\tau) \right] b(\tau) \qquad (6.2)$$

$$L(T) = \sum_{\tau=1}^{T} \left[I(\tau) - S(T,\tau) \right] \frac{b(\tau)}{a(\tau)} \qquad (6.3)$$

where τ is the year in which a particular vintage of machinery, itself characterized by the label τ, comes onstream. It is assumed that the oldest vintages are scrapped first. The above procedure ensures that the transition from 'hypothetical' data is chosen such that any increase in that length produces no significant change in the results obtained.

The second phase covers the period from the year T to that for which most recent data are available – in the empirical applications discussed in the next chapters, this phase covers 1954–83. During this period the implied technical coefficients of newly installed vintages are calculated from real data, as described below.

Using published investment and capital stock estimates as input, the quantity of capital scrapped in year t is given by

$$S(t) = K(t-1) - K(t) + I(t). \qquad (6.4)$$

Some assumption is needed whereby this scrapped capital is allocated among vintages so that the productive and employment potential 'lost' by this scrapping can be estimated.

All or part of the oldest 'vintage' of capital which is assumed to have been operating last year is retired in the current year, and the corresponding output and labour 'losses' are estimated from the corresponding marginal output:capital ratios. Any outstanding scrapped capital is transferred to the succeeding vintage and the procedure repeated until all the $S(t)$ has been accounted for. If $Q^s(t)$ and $L^s(t)$ are potential production and employment 'lost' in year t, then the contributions of surviving vintages are:

$$\overline{Q}(t) = Q(t-1) - Q^s(t)$$
$$\overline{L}(t) = L(t-1) - L^s(t). \qquad (6.5)$$

The technical coefficients associated with the vintage installed in year t can now be estimated. The value of $\bar{Q}(t)$ and $\bar{L}(t)$ obtained by this procedure are subtracted from the actual values of output Q and employment L in the year concerned to give the marginal contributions ΔQ and ΔL which are associated with equipment brought onstream in the year under consideration. Hence the first approximations to the contribution to total output and employment from equipment installed in year t are calculated as

$$\Delta Q(t) = Q(t) - \bar{Q}(t)$$

$$\Delta L(t) = L(t) - \bar{L}(t)$$

(6.6)

respectively, with $\bar{Q}(t)$ and $\bar{L}(t)$ given by equations (6.5). Then a first approximation to the output:capital and labour:capital ratios associated with this equipment can be formed as $\Delta Q(t)/I(t)$ and $\Delta L(t)/I(t)$.

It is clear that by no means all the variations in $\Delta Q/I$ and $\Delta L/I$ are caused by year-to-year changes in the nature of the technology embodied in the newly installed capital equipment. 'Labour hoarding', 'shake-outs' and numerous other phenomena can have a significant, often dominant, effect on the marginal ratios calculated as above.

A smoothing procedure is therefore adopted, as follows. Given observations for 36 years (1948–83 inclusive) the logarithmically smoothed value at time t (where year 1948 corresponds to $t=0$) of the aggregate output:capital ratio is given by

$$\left(\frac{\bar{Q}}{K}\right)_t = \left(\frac{\bar{Q}}{K}\right)_{1948} \exp(\lambda t)$$

(6.7)

where $\lambda = \dfrac{1}{35} \ln \dfrac{(Q/K)_{1983}}{(Q/K)_{1948}}$

The output:capital coefficient $b(\tau)$ associated with equipment installed in year τ is then calculated as

$$b(\tau) = b(\tau - 1) + \left[\left(\frac{\bar{Q}}{K}\right)_t - \left(\frac{\bar{Q}}{K}\right)_{t-1}\right] + \alpha \left[\frac{\Delta Q(t)}{I(t)} - \left(\frac{\bar{Q}}{K}\right)_t\right]$$

(6.8)

If $\alpha = 0$, the $b(\tau)$ follow a path corresponding to that of the smoothed aggregate coefficient; the implication of this is that observed fluctuations in that coefficient around the long-term trend merely represent changes in capital utilization rates, and that the technical characteristics of successive generations of equipment change smoothly. If $\alpha > 0$, the $b(\tau)$ will fluctuate in response to variations in $\Delta Q/I$ about the trend. For example, a large increase in output without a correspondingly large increase in investment will show up partly as an above trend increase in $b(\tau)$ and partly as an increase in capital utilization.

The use of the above equation with a nonzero α (and of a similar equation for the labour–capital coefficient $c(\tau)$, from which $a(\tau)$ can be derived, or vice versa) is intended to capture the notion that, at least in the longer run, technical change is not a smooth process; while year-to-year fluctuations in the coefficients have little meaning, it is assumed that variations from trend in the ratios $\Delta Q/I$ and $\Delta L/I$ over longer periods represent technical phenomena and not merely variations in utilization rates. We argue that the results of the analysis provide some support for this contention. In addition, the ability of the model to trace the effects of changes in technology is useful for forecasting purposes.

This procedure is carried out for each successive year from T to the most recent for which data are available, yielding a notional picture of

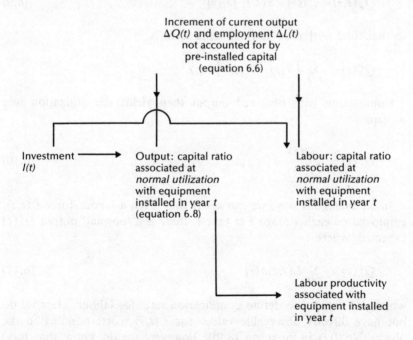

Figure 6.2 The vintage model flow chart

the technologies embodied in the capital stock. From the estimated marginal coefficients and the aggregate data, utilization rates can be inferred, as discussed shortly. Figure 6.2 shows a flowchart summarizing the steps involved.

The third phase involves switching the model into a 'forecasting' mode by converting some of the previously endogenous variables into exogenous variables, and vice versa. The inputs are now assumed values of future marginal technical coefficients, output, labour utilization, investment and scrapping rates. From these, aggregate values can be obtained for employment, capital utilization (and capital stock) and average ratios. Of particular interest for current purposes is the use of the model to speculate about what combinations of increased demand and increased supply capacity (through new fixed investment) may be required to effect significant reductions in unemployment.

The utilization rates for capital and labour are calculated as follows, given the 'technological' marginal output:labour ratios $a(\tau)$ and output: capital ratios $b(\tau)$, associated with successive vintages of equipment τ and estimated as above. The 'potential' or 'normal' output from vintage τ is

$$Q_K^P(t,\tau) = \left[\, I(\tau) - S(t,\tau) \,\right] b(\tau) \tag{6.9}$$

so that total output at 'normal' utilization is

$$Q_K^P(t) = \sum_{\tau=1}^{t} \left[\, I(\tau) - S(t,\tau) \,\right] b(\tau).$$

Comparison with observed output then yields the utilization rate of capital

$$U_K(t) = \frac{Q(t)}{Q_K^P(t).} \tag{6.10}$$

In an analogous way, we can assert that given a labour force $L(t,\tau)$, employed on each vintage τ at time t, there is a 'normal' output $Q_L^P(t)$ expected, where

$$Q_L^P(t) = \sum_{\tau=1}^{t} L(t,\tau) a(\tau) \tag{6.11}$$

which can be used to define a 'utilization rate' for labour. Here we do not have directly observable values for $L(t,\tau)$ (corresponding to the observable $I(\tau)$ in equation (6.9)); however, we do know that total employment $L(t) = \Sigma_\tau L(t,\tau)$, where the sum is over surviving vintages, just as the capital stock $K(t) = \Sigma_\tau [I(\tau) - S(t,\tau)]$. We also have a value for 'expected' labour force employed on vintage τ, namely

$$\epsilon(t,\tau) = \left[I(\tau) - S(t,\tau) \right] c(\tau) \tag{6.12}$$

where $c(\tau)$ is the labour:capital ratio associated with vintage τ, i.e.

$$\epsilon(t,\tau) = Q_K^P(t,\tau)/a(\tau) \tag{6.13}$$

from equation (6.9).

Now making the assumption that the ratio of 'expected' to 'actual' employment is the same across all vintages, we have

$$\frac{\epsilon(t,\tau)}{L(t,\tau)} = \frac{L_0(t)}{L(t)} \tag{6.14}$$

where

$$L_0(t) = \sum_{\tau=1}^{t} \epsilon (t,\tau). \tag{6.15}$$

The ratio in equation (6.14) could be taken as a measure of labour utilization, the ratio exceeding unity if 'actual' employment is less than the 'expected' (implying that the employed labour is working at above normal efficiency given the nature of embodied technology), and vice versa. However, this measure does not allow for variations in labour utilization if output fluctuates strongly, other things remaining equal (since output does not appear in equation (6.12)), and is hence not exactly analogous to the definition of capital utilization adopted. It is therefore preferred to return to equation (6.11) which, with equation (6.14) gives

$$Q_L^P(t) = \sum_{\tau=1}^{t} \epsilon (t,\tau) \frac{L(t)}{L_0(t)} a(\tau) = \sum_{\tau=1}^{t} Q_K^P(t,\tau) \frac{L(t)}{L_0(t)} \tag{6.16}$$

from equation (6.13). Thus

$$Q_L^P(t) = Q_K^P(t) \cdot \frac{L(t)}{L_0(t)} \tag{6.17}$$

Now the utilization of labour can be defined as

$$U_L(t) = \frac{Q(t)}{Q_L^P(t)} = \frac{L_0(t)}{L(t)} \cdot U_K(t) \tag{6.18}$$

where $L_0(t)$ is found from equations (6.15) and (6.12). Equation (6.18) may alternatively be written

$$U_L(t) = \frac{Q(t)}{L(t)} \frac{\Sigma_\tau \ [I(\tau) - S(t,\tau)] \ c(\tau)}{\Sigma_\tau \ [I(\tau) - S(t,\tau)] \ b(\tau)}$$

i.e. the average productivity of labour $A(t)$ is

$$A(t) = \frac{\Sigma_\tau \ [I(\tau) - S(t,\tau)] \ b(\tau) \ U_L(t)}{\Sigma_\tau \ [I(\tau) - S(t,\tau)] \ b(\tau)/a(\tau)}$$

since $c(\tau) = b(\tau)/a(\tau)$. Hence, for example, an observed rise in average productivity of labour may be reflected partially in a rise in the utilization rate U_L and partially in a rise in the marginal ratio $a(\tau)$. Whether the ratio $a(\tau):A(t)$ rises or falls depends on 'how much' of the variation in $A(t)$ is accounted for by change in the marginal ratio, as determined by the method discussed above. The ratio $a(\tau):A(t)$ provides a measure of the gap between the 'frontier' techniques and the industry average practice. Along with the other calculated quantities, this ratio can vary with time in a discontinuous way, assumptions such as smooth trends or linear relationships between variables being avoided in this approach.

To conclude this section, it should be pointed out that a good deal of uncertainty exists in several of the data and assumptions used, in particular, with regard to the following:

(1) The input data on the capital stock. This is taken from Armstrong (1976) and updated, using official series published in *National Income and Expenditure* (1983). The method used by Armstrong is the 'Perpetual Inventory Method' which is based on the assumption that each of the particular categories of capital equipment has a fixed lifetime, independent of technological or economic developments (for a discussion and defence of the methodology see Griffin (1976)). The resulting series for the capital stock (the series for 'plant and machinery' in fact has been used here) is widely regarded as subject to substantial error, and in particular to represent an overestimate of the stock. To correct for this overestimation, which in many sectors looked particularly acute for the period after 1973, the capital stock series was re-estimated in order to allow for the scrapping of physically 'productive' but economically 'unproductive' capital. The scrapping derived in equation (6.4) may be regarded as the physical obsolescence of capital stock. An additional term based on the level of under-utilized

capacity, as derived from the model, can be incorporated into this equation. In effect this term allows for the additional scrapping of economically unproductive capital. The rationale behind the 'additional scrapping criterion' is that capacity utilization is a short-term cyclical variable and any long-term trend towards lower utilization, a concomitant of overestimated capital stock, seems suspicious and is not borne out by the CBI industrial trends survey. Thus, in the re-estimation, 'increased' scrapping occurs when capacity utilization remains below certain prescribed levels for more than 5 years. This effectively reduces capital stock levels and stems any long-term decline in capacity utilization.

(2) Delays between investment expenditures and the coming onstream of new productive capacity. It is clear that such delays can be highly variable, from very small in industries where new plant is purchased chiefly 'off the peg' (e.g. some textiles), to perhaps several years in capital-intensive industries where construction lead times are long and payments may begin long before the plant is available for operation, e.g. chemicals. In any industry the published investment expenditure estimates for a particular year will include payments for equipment beginning operation in the year in question, that not due to be used until some time in the future, and also (probably) some delayed payment for plant already in use. A further difficulty relates to the increasingly important question of leasing of equipment, where investment expenditures are undertaken by one sector but the equipment used by another. Again the only way of treating this is to assess sensitivity to alternative assumptions within the range of plausibility or the industry under consideration.

(3) Having obtained the apparent capital and labour productivities from the scrapping assumptions, the derived technical coefficients of retired vintages (giving Q^s and L^s) and the estimate of the value of productive capacity coming onstream in the current period, the seemingly arbitrary way in which the separation between long-term ('embodied') and short-term fluctuations is effected, needs to be defended. First, it should be stressed that the main purpose of this analysis is only to capture major trends in rates of technical advance over several years, rather than striving for precise year-to-year estimates. It also turns out that in many cases, the qualitative results obtained are insensitive to the detailed assumptions regarding the 'allowable' change in the technological ratios. A further check is that other estimates of utilization rates are available for comparison and also that one would not, in general, expect these rates which, by assumption, are due to variations in short-term factors, to show any long-term secular trend.

98 J. A. Clark

It is certainly not claimed that we have overcome all the limitations of modelling these complex systems. It is evident from this discussion that we are well aware of some of the main limitations of our own model. We agree with Schumpeter that there are some inherent limitations to the use of such models when qualitative changes are taking place. We claim only that this simulation model offers a somewhat more realistic representation of the behaviour of some parts of the system than other models which have been widely used.

7
Future Employment Trends in UK Manufacturing Using a Capital-Vintage Simulation Model

John Clark, Pari Patel and Luc Soete

In the following pages, using the vintage model set out in the previous chapter, we shall concentrate on the effects of the adoption and transmission of new techniques into the production system of the UK manufacturing sector via fixed capital formation.

It is clear that new technology affects investment and employment in two ways. First, it generates changes in the scale and direction of new investment by offering new opportunities for profitable exploitation of new and improved products and processes and in response to competitive pressures. Secondly, the newly installed 'vintages' of equipment offer employment opportunities, which vary with the embodied technology. The questions on which we wish to throw some light include: (1) Is UK manufacturing employment growth likely to be constrained by an inadequate supply of fixed capital, taking account of the technical characteristics of new equipment? (2) What levels of output and investment might be necessary to maintain or increase UK manufacturing employment?

As discussed in chapter 6, the fundamental assumption behind our vintage capital model is that it is possible to associate with capital equipment coming onstream in a given year a particular labour input requirement and a particular level of output at 'normal' utilization. These technical characteristics are then assumed to be retained by the equipment throughout its working life. The model lends itself *par excellence* to the manufacturing sector. It should be remembered that this sector accounted

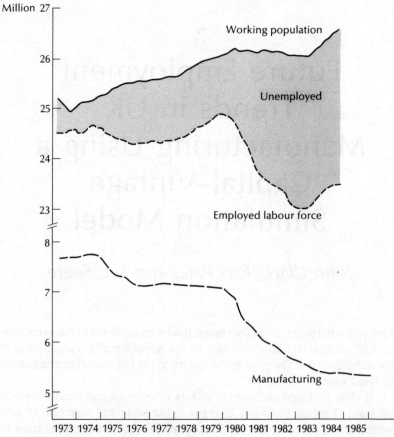

Figure 7.1 Working population, employed labour force, manufacturing employees in employment (Great Britain seasonally adjusted 1973–85)

for the largest part (some 1.8 million) of the 2 million jobs lost in the United Kingdom over the period 1979–83 (figure 7.1). Even over the last two recovery years, employment in manufacturing continued to fall.

Structural Unemployment: Some of Issues Definition

Fluctuations in employment may in principle occur through fluctuations in the rate of investment and/or through mismatches developing between the capital stock, the available labour force and its skills, the changing pattern of demand and the changing potential of new technology. Since in practice there are always imperfections in the flow of information about employment opportunities as well as in the mobility of labour and its adaptability, there will probably also always be some level of 'frictional' unemployment, even when demand for labour is strong. Whilst there is no precise definition, 'structural' unemployment is generally defined as that type of unemployment which arises when the time-lags involved in the adjustment process are considerable, as for example in the acquisition of entirely new skills, or the replacement in a particular region of an obsolete vintage of capital equipment in a declining industry by new types of capital stock in a newly expanding industry.

Structural unemployment results both from structural 'labour mismatches', such as regional and skill mismatches and 'capital' mismatches, sometimes referred to as 'capital shortage' unemployment (OECD, 1983)[1] where unemployment appears to be the result of an apparent lack of 'productive' capital, whether because of lack of capital–labour substitution possibilities, rigidities in relative input prices, or sluggishness of investment behaviour in particular industries, regions or countries. We prefer to use the term 'capital mismatch' unemployment, which puts the emphasis more directly on the side-by-side coexistence of excess capacity and capital shortage in the economy at large, as well as within specific industrial sectors.

From a practical point of view there is, however, no easy way to separate out cyclical unemployment from structural unemployment. The distinctions between cyclical or demand-deficient, structural and frictional unemployment, while analytically useful and particularly relevant from a policy perspective, raise many empirical questions and are difficult to quantify, each category overlapping with the other to a significant extent. In terms of neo-classical analysis, all three categories of unemployment are related to lack of instantaneous price adjustments, focusing each on a different facet of the adjustment lag: in the case of demand-deficient unemployment, lack of instantaneous competitive price adjustment in factor and output markets; in the case of frictional and skill mismatch unemployment, lack of instantaneous wage and skill-related wage-differentials adjustment; and in the case of structural unemployment, rigidity of relative input prices and lack of instantaneous capital–labour substitution.

There is also no stability in each of these unemployment categories; an increase in one category of unemployment will frequently lead to an increase in the other. Thus it could well be argued that the present recession has initially led to an increase in demand-deficient unemployment which, however, with the induced slowdown in investment, has gradually taken the form of an increase in structural unemployment as firms adjust their capacity to the lower expected output levels through the scrapping of older vintages of capital equipment.

102   J. A. Clark, P. Patel and L. Soete

But whatever the difficulties involved in separating out the various unemployment components, some assessment of the importance of the three categories of unemployment, and particularly of cyclical versus structural unemployment, seems essential for one's understanding of the scale and the nature of the present unemployment problem and as a guide to the choice of policy instruments. The most straightforward way of separating out cyclical unemployment from structural unemployment relates to the so-called 'Okun curve', relating capacity utilization to

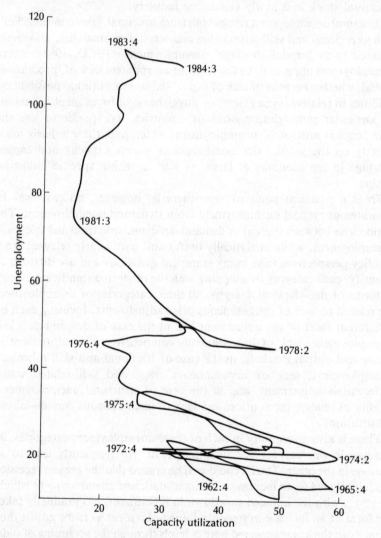

Figure 7.2a Okun curve for UK

unemployment. Figure 7.2a illustrates this relationship for the United Kingdom on the basis of quarterly data. Unemployment rate is plotted along the vertical axis and capacity utilization along the horizontal axis. At a 'normal' degree of capacity utilization demand-deficient

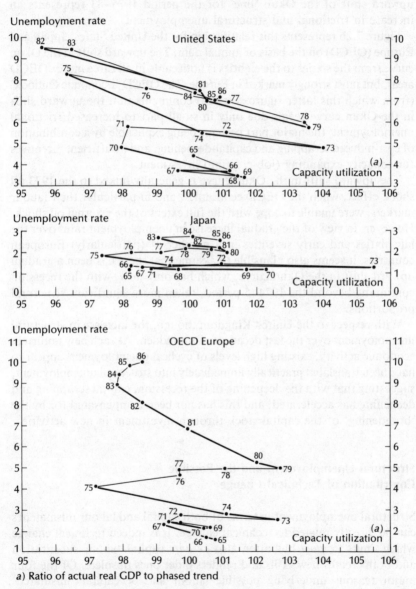

Figure 7.2b Okun curves for the USA, Japan and Europe
Source: OECD, *Economic Outlook*, June 1985

a) Ratio of actual real GDP to phased trend

unemployment can reasonably be assumed to be insignificant, the existing unemployment relating primarily to frictional and structural unemployment. Demand-deficient unemployment is represented by a movement to the left along the Okun 'line' as in the 1975 and 1982–3 recessions. However, the upward shift of the Okun 'line' for the period 1975–83 represents an increase in frictional and structural unemployment.

Figure 7.2b represents this relationship for the United States, Japan and Europe (OECD) on the basis of annual data. The upward shift in the Okun curve from the sixties to the eighties is noticeable in all three major OECD areas, but most strongly marked in Europe. The OECD 'Economic Outlook' (from which this latter figure is taken) comments that the upward shift in the Okun curve seems due only in small part to increased frictional unemployment, the major part perhaps being explicable by a combination of cost-induced scrapping and capital-deepening, and insufficient incentives for capacity-expanding (job-creating) investment.

The upward shift in the Okun curve is generally linked to the 1973 oil shock effect, when the OECD economies, and in particular their labour markets, were unable to cope with the full extent of the external oil shock. However, in view of the gradual increase in unemployment rates over the late sixties and early seventies in a number of (particularly) European countries, it seems also plausible to assume that there has been a gradual upward shift in the Okun curve, which has only now, with the recession continuing beyond the 1974–5 period, taken more significant structural proportions.

With respect to the United Kingdom though, the increase in structural unemployment over the last decade is more sudden. At each new upturn in economic activity, existing high levels of cyclical unemployment appear to have been translated practically immediately into structural unemployment, suggesting that with the deepening of the recession, capital scrapping and deepening has accelerated, and this has not been compensated for by the 'broadening' of the capital stock through investment in new activities.

Structural Unemployment and the Possible Contribution of Technical Change

Structural unemployment in terms of both capital and labour mismatches can be directly related to technical change. It is indeed technical change which limits *ex post* but often also *ex ante* capital/labour substitution and both creates new skills and renders older ones obsolete. Of the four major reasons underlying possible labour and/or capital 'mismatch' unemployment – a slowdown in investment; a tendency towards rationalization investment; acceleration in scrapping; and increased new skill requirements – the last three can more or less be related directly to the rate and direction of technical change, with the possibility of a tendency towards labour-saving technical change rendering both large parts of existing physical and 'human' capital obsolete. As we have seen from Lederer's thesis advanced in chapter 2, even the first one – the slowdown in investment over the post-1973–5 period – could well be indirectly, and maybe paradoxically, related to the increased speed in the emergence of a set of radical new technologies such as information technology. It could be argued that the diffusion of these new technologies is as much hampered by the overall recessional climate as by the radical nature of the new technology itself: its 'immature' and unstandardized nature; its widespread cross-industry area of application; the lack of reliable, easily accessible information on the profitability of the new technology within the specific user's environment; cash-flow problems; and, most of all, the shortage of the new skills needed. These might all provide powerful retardation factors, leading firms to postpone investment decisions until the technology has matured, new skills become available and the investment is less risky (for an attempt to model this see Soete and Turner, 1984).

There is no clear-cut way in which this assumed impact of new technologies on the slowdown in investment could be reasonably quantified although some empirical evidence is reported in our sector studies. The best one can do is to point to technical change as one of the major 'uncertainties' determining investment decisions.

However, using the capital–vintage model set out in the previous methodological chapter it will be possible to look somewhat closer at the 'assumed' contribution of the underlying rate of technical change. We first discuss capital productivity.

Capital Productivity

Curve A in figure 7.3 shows the aggregate capital productivity derived directly from the output and capital stock data input to the model. These data indicate that (apart from cyclical variations) the productivity of plant and machinery in the United Kingdom declined progressively from the mid-fifties to the early eighties. The direct implication of this is that new technologies introduced over this period were intended specifically to save labour costs, and that this was achieved at the expense of increased capital costs per unit of output. The trends in labour productivity over this period are discussed later.

Curve B in figure 7.3 shows the apparent trends in the productivity of *newly installed* capital equipment, as derived from the model discussed in chapter 6. It is obtained from equation (6.8) with a small positive coefficient in the final term. This shows that the productivity of such

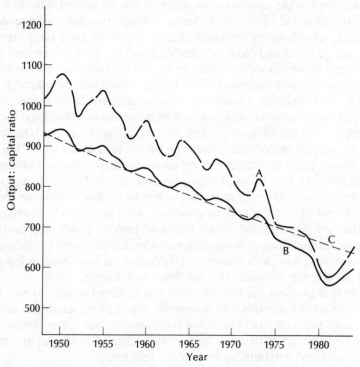

Figure 7.3 Average and best practice capital productivity

equipment 'leads' the aggregate values, i.e., since the aggregate values are weighted sums of historical 'best-practice' values, and since the overall trend is downward, one would expect the best-practice values to be, in general, lower than the aggregates in any particular year. The straight line (marked C) shows the $b(\tau)$ with the final term in equation (6.8) set equal to zero. This results from the assumption that all fluctuations in the aggregate arise from variations in capacity utilization – the trend in embodied technology is taken to be linear. As can be seen, any allowance for the trend in average productivity to be reflected in 'best-practice' productivity (as in curve B) suggests that the rate of decline in the $b(\tau)$ may have accelerated during the early seventies; such a modification in the direction of technological change can have important implications for the levels of investment required for a return to a full-employment path, as will be discussed later.

Figure 7.4 shows the corresponding curves for capacity utilization, with the same assumptions used to obtain curves B and C in figure 7.3. As expected, the variations in utilization are somewhat greater where no variation is allowed in best-practice capital productivity (curve C) – in this

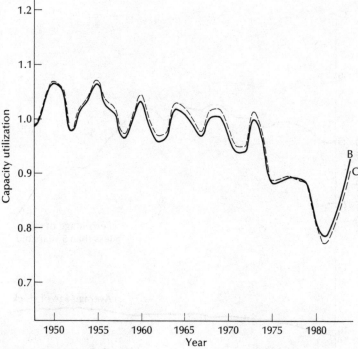

Figure 7.4 Capacity utilization

case all fluctuations in output are translated into fluctuations in utilization. The pattern is, however, not very sensitive to the assumptions used, both curves showing the expected cyclical pattern of utilization, with a peak in 1973, significant declines in 1974 and 1979 and recovery in the early eighties.

The Quarterly Industrial Trends Survey conducted by the Confederation of British Industry (CBI) reports the percentage of firms surveyed operating at below capacity. The rise in utilization 1975–9 implied by these data is more pronounced than the model suggests, but this evidence supports the overall decline 1972–80[2] and the rise in the early eighties. The important implication for employment is that the industrial capital base is now such that, even if an upsurge in demand were sufficient to restore 'full' utilization, little increase in labour demand could be expected. Any significant improvement in industrial employment prospects depends not only on an upsurge of demand for the output of the sector, but also on its ability to supply. Unless there is a large upswing in investment, or an increasing trend towards 'capital saving' technologies, a severe supply-side constraint to higher employment seems probable. Later in this chapter we examine this issue further.

Figure 7.5 Age of the capital stock

Before leaving the discussion of features associated with the capital stock, it is interesting to look at changes in the age structure of plant and equipment over time. Two measures are illustrated in figure 7.5: the average age of the stock in use, and the percentage of the total installed over the preceding 5 years. While there are small variations between the alternative model runs, it appears that the average age of stock has remained roughly constant (at 11–12 years) since the late fifties. The proportion of stock in operation for 5 years or less, on the other hand, seems to have declined from over 27 per cent in the mid-sixties to around 20 per cent in the early eighties. Taken together, these results are consistent with a profile of the stock with increased bunching around the middle-age ranges, i.e. lower growth in investment (giving less young equipment) combined with increased decommissioning of old stock.

Labour Productivity

The capital–vintage model also produces estimates of the output–labour

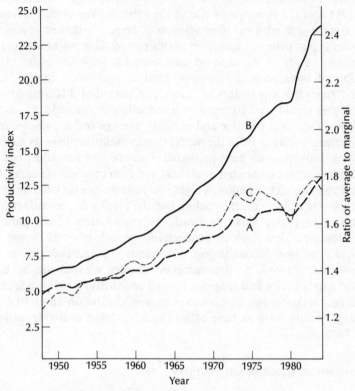

Figure 7.6 Average and best practice labour productivity

ratios $a(\tau)$ associated with successive vintages τ of equipment in a similar way to the $b(\tau)$. Figure 7.6 shows the overall measured labour productivity (output per employee-year, curve A) together with the estimated $a(\tau)$ (curve B). These results were produced on the version of the model corresponding to curve B in figures 7.3 and 7.4. Curve C shows a ratio R_L formed by dividing the $a(\tau)$ by the measured output per employee year 'corrected' for variations in utilization rates; the denominator therefore represents the notional overall productivity at 'normal' utilization of the factors of production. Hence the ratio provides a measure of the extent to which the labour productivity associated with new equipment exceeds that of all equipment in use, i.e. the degree to which 'frontier' technologies are more efficient, in terms of output obtainable per unit of labour, than the manufacturing average.

It turns out that the values of $a(\tau)$, and hence also of the ratio described above, are sensitive to the precise assumptions used in the model. However, the general form illustrated in the figure shows features which are typical of a number of runs of the model. In particular, rapid growth in the $a(\tau)$ after 1965 and the consequent rise in the ratio R_L are readily obtained and are consistent with the phenomenon of large economies of scale of new process plant coming onstream in this period. One might also expect high values of R_L in the early to mid-seventies, with the onset of the adoption of labour-saving computer control equipment in the 'best-practice' firms in many sectors of industry. A period of diffusion of such technologies would tend to result in a reduction in the ratio R_L, as the gap between the best-practice and industry average techniques declines. There is some evidence from the model that in manufacturing as a whole the gap at least may not have increased since the late seventies, but the results obtained are somewhat equivocal and firm conclusions cannot be drawn. From our sector studies, a highly variable picture between different industries was evident on this point. But the results for manufacturing industry in aggregate illustrate clearly an explanation of the paradox of the observed slowdown in productivity growth after 1973 and the availability of new labour-saving microelectronic technologies in the late seventies. There is no inconsistency between a rapid rise in 'best-practice' productivity and stagnant overall productivity – slow adoption of the new technologies combined with low utilization rates ('labour hoarding') would seem to have offset the advantages available, at least up to 1980.

Employment–Output Relationships

As a precursor to an examination of possible employment forecasts we now turn to consider how employment in manufacturing industry has been related to output growth over the postwar period. Previous work (Clark, 1983) presented the results of estimating a simpler linear relationship between output growth and productivity growth (the 'Verdoorn' equation) for UK manufacturing 1949–79. For each successive rolling 10-year period (1949–58; 1950–9, etc.) the equation

$$\frac{\dot{P}}{P} = A + B\frac{\dot{Q}}{Q} \tag{7.1}$$

was estimated using ordinary least-squares regression, where P is productivity (output per employee-year), Q is net output and the dot denotes differentiation with respect to time. From the values of A and B obtained, the identity

$$\frac{\dot{P}}{P} = \frac{\dot{Q}}{Q} - \frac{\dot{L}}{L} \tag{7.2}$$

(where L is employment) can be used to find the value of output growth which, in each period, is just sufficient to maintain employment at a constant level (i.e. with $\dot{L} = 0$); this value is given by $\lambda = A/(1 - B)$.

The results suggested strongly that the structure of equation (7.1) had changed substantially over the postwar period, in such a way that employment creation became increasingly 'more difficult' in the sense of requiring an increasingly rapid rate of output growth – with minor fluctuations, the parameter λ grew steadily over successive 10-year periods. The values of λ are, of course, dependent on the length of the time intervals chosen, but alternative choices appeared to confirm the increasing trend.

The existence of a statistically significant relationship of the form of equation (7.1) for any particular period depends, at least partly, on the cyclical behaviour of the economy; it is not surprising that productivity, output and employment should tend to move together in alternating short-term booms and recessions. We can, however, use the model described in chapter 6 to assess whether there is any relationship between these variables when cyclical variations are removed, using the 'expected' (normal utilization) value of employment $L_0(t)$ and the definition of potential output $Q^P(t)$ (see chapter 6). As weighted averages over all generations of capital in use, these variables move far more smoothly than the marginal coefficients $a(t)$ and $b(t)$, and variations are much less sensitive to the particular assumptions used in the model.

In figure 7.7 year-on-year growth rates of $L_0(t)$ are plotted against those of $Q^P(t)$ for a 'typical' model run – the results are broadly independent of the specific version chosen. It is noteworthy that up to 1966 the points are clustered in the upper right-hand quadrant, with underlying (potential) output growth being consistently associated with positive underlying employment growth. The onset of employment decline after 1966 corresponds to an apparent decline in the growth rate of $L_0(t)$, while the growth rate of $Q^P(t)$ remains approximately constant up to 1970, and declines steadily thereafter. The periods 1948–66 and 1971–83 appear to form two distinct epochs in the relationship between these variables; in particular, a lower underlying average output growth is required to maintain $L_0(t)$ constant in the earlier period (the value is about 2 per cent, as given by the intercept of the estimated regression line with the horizontal axis). For the later period the required growth rate is of the order of 3.5 per cent. This tendency is in line with the results of the earlier analysis.

The results can be interpreted as showing a change in the nature of new technologies incorporated in the production system in the sixties and seventies, away from those designed primarily to allow increased output and towards those which are predominantly labour saving. Other evidence

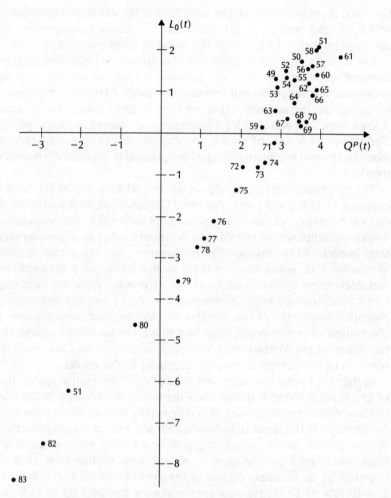

Figure 7.7 Percentage growth rates of potential output and employment

for a shift from 'expansionary' to 'rationalizing' investment has been
discussed by Freeman (1979).

Manufacturing Employment 'Futures'

Is there a capital constraint?

Evidence presented earlier in this chapter suggests that the amount of
unused capacity in the manufacturing sector has declined in recent years
despite high and increasing levels of unemployment. A sustained period

of low demand and high unemployment might be expected to be accompanied by a reduction in the level of productive capacity, after which any upturn in demand could fail to reduce unemployment owing to a supply-side constraint. The model described here can, however, be used to examine the conditions under which such a problem could arise. The question can be put in terms of the investment and output levels required to maintain manufacturing employment at its present level. The conclusions reached depend on the rate at which existing equipment is scrapped and the output: capital and output:labour ratios associated with newly installed capital.

First, we assume that the value of capital utilization tends asymptotically to the 'normal' value of unity, reaching 97 per cent of this level by 1990. In addition, we project forward past historical trends in the technical coefficients associated with new equipment and use a scrapping rate of 2 per cent per annum, roughly in line with the average observed value since 1970. Under these conditions, the model suggests that in order to maintain manufacturing employment at its present level an average output growth of around 3 per cent would be required. However, in order to achieve that output growth, an average investment growth of around 10 per cent a year is required over the period 1982–2000; nearly ten times the historical average investment growth of just above 1 per cent a year. The severity

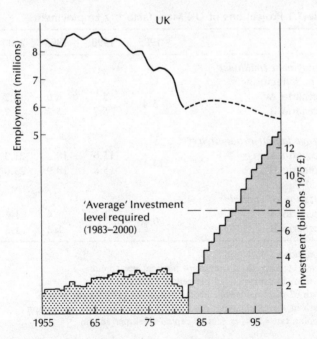

Figure 7.8 Investment requirements and employment forecasts 1955–2000, UK manufacturing

of the capital bottleneck for future output growth in the UK manufacturing sector is illustrated in figure 7.8. Particularly in the short- to medium-term (1982–90), the required investment growth seems particularly high – more than 14 per cent a year. As illustrated elsewhere (Soete and Freeman, 1985), similar but less severe problems seem also to confront the other European countries, with relatively high output growth rates needed to maintain employment at present levels, and 'required' investment growth rates above the average historical investment growth rates. This contrasted sharply with the case of the United States and Japan, where no evidence of any capital shortage in manufacturing emerged.

The results of simulations with a more modest growth rate in investment, but with a growth rate of output sufficient to yield a gradual increase in employment, are shown in table 7.1. Apart from the output growth assumption and the correspondingly optimistic assumption regarding growth of fixed investment, scenario A corresponds broadly to the pattern observed since 1970. Scenario B is identical except in the assumption of a reversal in the historical trend of declining capital productivity.

Comparison of the implications for the utilization of capital shows that scenario A leads to unprecedentedly high utilization requirements during

Table 7.1 Projections of UK Manufacturing Employment

	1985	1990	1995	2000
Employment (millions):				
UK manufacturing				
Scenario A	5.4	5.7	6.0	6.2
Scenario B		5.7	5.7	5.7
Average labour productivity:				
Scenario A	13.4	15.6	18.2	21.3
Scenario B		15.8	18.9	23.0
Utilization of capital:				
Scenario A	0.96	1.2	1.4	1.6
Scenario B		1.1	1.2	1.2

Assumptions:
Scenario A: Growth of output = 4 per cent per annum
Growth of investment = 2.5 per cent per annum
Growth of marginal labour productivity = 2.5 per cent per annum
Growth of marginal capital productivity = − 1.1 per cent per annum
Scrapping rate = 2.5 per cent of capital stock per annum

Scenario B: As scenario A, except that growth of marginal capital productivity + 5 per cent per annum.

the nineties, implying that a supply-side constraint would operate to prevent the output growth assumed from being achieved in practice. This constraint is far less marked in scenario B, where the rather dramatic reversal in the trend of technology associated with new equipment towards capital saving is sufficient to significantly reduce this problem. In this latter scenario, the higher output–capital ratios compared with scenario A, combined with the same marginal output–labour ratios, imply higher labour–capital ratios associated with new equipment. Hence the 'expected' employment $L_0(t)$ is higher in this scenario; the fact that the projection shows a lower actual employment path (table 7.1) arises because this effect is, in this case, more than offset by the lower utilization rate. In both examples, the recovery in employment is small despite the high growth rate assumed – employment in manufacturing does not return to even the relatively low levels of the late seventies by the end of the century.

The overall conclusion from the analysis presented here is that a capital constraint to growth could well occur in the United Kingdom in the medium term (late eighties and early nineties), but could be overcome by a combination of higher capital productivity associated with new technology and of higher levels of investment in manufacturing.

Can manufacturing employment return to the levels of the sixties or seventies?

The results already presented suggest that only a modest rise in employment in manufacturing can be expected even under the assumption of a high rate of growth of output. Here some evidence is presented of the possible effects on employment of a return to economic conditions prevailing in earlier decades, and the requirements for a substantial rise in manufacturing employment are assessed.

First, we assumed conditions approximating those of the sixties, with manufacturing output growing, on average, at about 2 per cent per annum and investment at about 5 per cent. Best-practice labour productivity grows steadily at 2 per cent, while best-practice capital productivity declines at an annual rate of slightly over 1 per cent. Secondly, we tested the implications of applying conditions similar to those experienced in the seventies, with the stagnant output and investment, rather lower growth in best-practice labour productivity (around 1.5 per cent per annum) and a slightly more rapid decline in best-practice capital productivity.

In both these scenarios, employment declines, in the latter case at rather more than 1 per cent per year, roughly comparable to the rate observed in the period 1966–73, but a much slower decline than that seen in the late seventies and early eighties. There is no suggestion in either case of a 'capital shortage' problem, in the first case because of relatively high

Table 7.2 Results using output and investment forecasts from the Institute for Employment Research, University of Warwick model

	Growth in:			Employment (millions)								
	Marginal labour productivity % p.a. (1)	Marginal capital productivity % p.a. (2)	Scrapping percentage of capital stock (3)	1982	1983	1984	1985	1986	1987	1988	1989	1990
A	2.5	−1.1	2.5	5.91	5.83	5.85	5.80	5.70	5.62	5.54	5.44	5.35
B	2.5	−1.1	1.7	5.91	5.88	5.93	5.93	5.8	5.85	5.80	5.74	5.67
C	5.0	−1.1	1.7	5.91	5.87	5.92	5.91	5.85	5.80	5.74	5.65	5.56
D	1.6	−1.4	2.0	5.91	5.86	5.90	5.89	5.83	5.78	5.73	5.66	5.59
E	2.5	−1.4	2.0	5.91	5.86	5.90	5.88	5.82	5.76	5.70	5.62	5.55
H	2.5	+5.0	2.5	5.91	5.83	5.85	5.79	5.68	5.58	5.49	5.38	5.26
					5.84	5.89	5.96	5.99	5.99	5.98	5.96	5.93

Warwick forecast	1982	1983	1984	1985	1986	1987	1988	1989	1990
Output	38.5	39.4	40.9	42.2	43.1	44.2	45.3	46.2	47.2
Investment	3.35	3.88	4.50	4.92	5.00	5.10	5.14	5.23	5.43

investment and in the second because, with constant output, the capital stock grows just rapidly enough to offset the effects of declining capital productivity.

A rise in manufacturing employment to around the 1980 level by 1995 appears to require an output growth rate of around 6 per cent per annum, together with a similar rate of growth in fixed investment and an annual rise in capital productivity of over 5 per cent to avoid severe problems associated with a capital constraint – an unlikely combination of events, suggesting again that we must look outside manufacturing for realistic prospects of substantial employment growth.

It is interesting to see in more detail how these generally rather gloomy employment prognoses depend on technical change for given future levels of output and investment. Values for output and investment from 1982 up to 1990 have been taken from forecasts produced by the demand-led macroeconomic model of the Institute for Employment Research, University of Warwick. These values are shown in the bottom right-hand corner of table 7.2 and were input to the model together with growth rates for the variables (1)–(3) in the table to produce the employment scenarios A–H. B corresponds to the 'sixties' conditions and D to the 'seventies' conditions for the three variables as used earlier; in the latter case in particular, however, output and investment growth are much higher here. Despite the optimistic path produced by the Warwick model for these variables, the results suggest a continuing downward path for manufacturing employment. The precise values obtained depend, of course, on the assumptions used for the best-practice technical coefficients and the scrapping rate; some of the choices are rather unlikely, for example the 2.5 per cent scrapping rate in A and H means that the capital stock grows very slowly by historical standards, despite average investment growth of 6 per cent, 1982–90. Overall, however, the range for manufacturing employment in 1990 implied by these scenarios is about 5.25–5.75 million – around two-thirds of the value 20 years earlier.

Also shown in table 7.2 is the employment path consistent with the given output and investment forecasts according to the Warwick model. As shown, the Warwick results suggest that manufacturing employment is about the same in 1990 as in 1982, with output rising at about 2 per cent per annum (on average) and investment rising by about 6 per cent per annum. Our results show a similar pattern – there is an overall decline in employment in each scenario, but it is not large except in those cases where scrapping is relatively high (2 per cent per annum, scenarios A and H).

On the whole, the employment forecasts are not sensitive to the growth rate of best-practice capital productivity assumed. This is, as previously discussed, important where problems of 'capital shortage' may arise, but this does not apply here, because of the high levels of investment assumed, at least within the time scale considered. Perhaps more surprisingly, they do not seem very sensitive to the growth in best-practice labour productivity – the reason is that even a step-function-like increase in this variable has little effect on overall employment until the corresponding capital has 'diffused' through industry, i.e. until heavy investment in the newer, highly productive generations of equipment has occurred. As pointed out above, the forecasts are rather sensitive to the scrapping rate – this is because of our assumption that the oldest plant, generally with the lowest associated labour productivity, is scrapped first. Lower scrapping thus implies more labour per unit of output overall, i.e. higher employment. As suggested earlier, however, an idea of the plausibility of the scrapping rate can be gained by comparing the capital stock with investment, which renders some of the scenarios in table 7.2 unlikely. From this and earlier analyses (table 7.1) the best central estimate of employment in UK manufacturing in 1990 is around 5.6 million, assuming moderately healthy growth rates.

The broad aggregate analysis presented here has obvious limitations. In the next, empirical, part we present the more disaggregated sectoral analysis covering both manufacturing and services. In part IV we return to the problem of estimating total future employment in the United Kingdom.

Notes

1 For the OECD, capital shortage unemployment 'reflects a seeming lack of "productive" capital to employ part of an "adequately" skilled and suitably located labour force, i.e. it is not accompanied by unfilled vacancies. Such a mismatch between available factor quantities may reflect limited *ex post* capital labour substitutability: once fixed investment has been installed, the amount of labour which is required to operate the capital is fixed and determines the demand for labour at full capacity utilization. But even if capital is malleable *ex post*, i.e. if labour and existing capital can be combined in varying proportions, the existing capital stock may be insufficient to provide enough jobs if labour costs relative to both other input costs and output prices fail to adjust sufficiently to absorb the available labour supply. Capital shortage unemployment can thus occur as a result of both lack of physical capacity and economic obsolescence: if variable costs exceed price, capital will not be used (and the corresponding jobs will disappear) even though such capacity could be operated in a physical sense' (OECD, 1983, p. 20).

2 The 1972–80 fall implied by the model would be even greater if the official series on the capital stock had been used directly. We are sympathetic to suggestions that the officially estimated capital stock series, calculated on the basis of the 'perpetual inventory' method, which assumes fixed lifetimes for particular categories of equipment, has in recent years yielded systematic overestimates of the stock. However, we have attempted to allow for this by increasing scrapping as utilization falls below certain levels.

III

EMPIRICAL ANALYSIS

In part II we discussed the use of mathematical models to illuminate and to project likely future trends in employment. Using a vintage capital simulation model suggests that significant growth in the UK manufacturing employment over the next 10 years is unlikely.

The detailed empirical analysis presented in the following chapters also focuses in the first instance on the United Kingdom. Elsewhere (Soete and Freeman, 1985; Soete, 1985a; Patel and Soete, 1985), analyses with regard to other OECD countries has been presented.

In the United Kingdom, the empirical debate about the likely employment implications of technical change came to the policy forefront somewhat before it did in other European countries. It saw its peak in 1978-9, with a variety of contributions from the trade union side (ASTMS, 1978; Sherman and Jenkins, 1978), the government side (DoE, 1979) and various research institutes alike (Freeman, 1979; Barron and Curnow, 1979; Maclean and Rush, 1979; Rothwell and Zegveld, 1979, etc.). This debate, with unemployment then running at 5-6 per cent, was by and large ignored by the academic economics community with only a few exceptions: most of the debate involved technology specialists and policy makers.

With the dramatic rise in unemployment, there has been a gradual increase of interest in the subject on the part of the economics profession. At the macroeconomic level, the Warwick Institute for Employment Research has been instrumental in broadening employment forecasts to include also the effects of new technologies, in particular microelectronics; at the micro-level, the Policy Studies Institute started some detailed case study research on the diffusion and use of microelectronics thoughout British Industry (Northcott and Rogers, 1982, 1985); finally the government itself, focusing increasingly on the diffusion of 'information technology' as one of its main industrial supply-side policies, has become increasingly concerned about employment skill shortages.

The UK debate is in many ways, however, an odd debate. If the prediction of Sherman and Jenkins of a 15 per cent rate of unemployment by 1985 appears indeed nearly fulfilled today, it remains nevertheless difficult to see how this 'successful' prediction could have anything to do with the 'sudden' introduction of microelectronics between the second quarter of 1980 and the fourth quarter of 1982 when unemployment more than doubled. The causes of unemployment in the United Kingdom are manifold; it would appear, though, that they are probably more closely related to lack of adequate productivity growth over a long period and loss of competitiveness, rather than to too fast a growth in labour productivity. It should therefore come as no real surprise if the rapid diffusion of microelectronics is often viewed in the United Kingdom as more positive in its employment compensation implications than in most other countries.

As already illustrated in the previous chapter (see figure 7.1) at the heart of the recent employment decline is the massive loss of jobs in UK manufacturing (some 42 per cent between 1979 and 1985). Not surprisingly, therefore, we start this empirical part with a detailed analysis of the various UK manufacturing subsectors in chapter 8. This analysis does show that the industries at the core of the new microelectronics technology were almost the only UK manufacturing subsectors to show increased employment in the recent period (1981–5). In chapter 9 the analysis turns to the UK service sectors and the postwar trends in technology and employment. Finally, in chapter 10, some empirical evidence of the emergence of a new information technology sector is provided.

Much of the analysis presented in this part concentrates on the issue of investment, capital mismatch and capital productivity and, following the previous chapter, the possibility of a capital constraint on employment growth in manufacturing. This should not be taken in any way as representing an underestimation of the problems of skill mismatch. On the contrary all the evidence of our sector studies as well as much other evidence from many different sources confirms that this problem is crucial in the United Kingdom and many other countries. Even though unemployment has been at very high levels throughout the eighties, there has simultaneously been a persistent and acute shortage of the key skills needed to implement information technology efficiently throughout the economy.

As in the analogous case of capital mismatch, the official figures do not reveal at first sight the true nature of the problem for several reasons. First, although the aggregate number of vacancies for such categories as computer engineers, systems designers, software engineers etc. is relatively small compared with the huge figures of total unemployed people, their importance is out of all proportion to their numbers. They provide the essential ingredient for introducing new applications for information technology and for expanding those manufacturing and service sectors which offer employment growth prospects. Secondly, the skill problem is not to be measured simply in terms of vacancies for highly skilled and qualified engineers and technicians. At all levels in the production and distribution system, there is a need for retraining of existing personnel in order to cope effectively with the change brought about by the new technology.

Finally, both in Britain and in other OECD countries virtually all studies (Freeman and Soete, 1985) of technical change and employment are agreed that the effective implementation of information technology over the next decade or two will require higher levels of computer literacy throughout the working population and a generally higher level of educational attainment. The British lag behind its principal competitors in educational and training provision is no less severe than the problem of capital shortage which has been described. We take up these policy implications in part IV.

8

Technological Trends and Employment in the UK Manufacturing Sectors

Pari Patel and Luc Soete

The UK manufacturing sector, it is worth remembering, contributes today less than a quarter of GDP. Twenty years ago the proportion was still just over a third. The decline in manufacturing's share of GDP has accelerated over the last decade with the onslaught of the recession which, particularly in the United Kingdom, appears to have vindicated the concept of 'de-industrialization' and, if anything, to have accelerated its trend. By international standards, the UK manufacturing sector is now a small sector, employing barely more people than e.g. the UK distribution service sector (SIC, Class 6).

Like most economic analyses, our focus on manufacturing, as already brought to the forefront in the methodological chapters, appears consequently somewhat out of proportion. The reasons traditionally invoked for paying particular attention to the manufacturing/industrial sector relate in the first instance to the international 'tradeability' of the sector, setting so to speak the balance of payment growth constraint for the rest of the economy. It is that argument which has been the subject of some recent debate in the United Kingdom following the House of Lords Select Committee Report (the so-called Aldington Committee, HMSO, 1985) on the decline in the international competitiveness of the United Kingdom (see a.o. Brittan (1985)). The extent to which services are being traded and might become more tradeable in the future is of course in the first instance an empirical debate which is hindered by the lack of official statistical data on the subject. We turn to this issue in our discussion of the impact of information technology in the service sector in chapter 10.

At the same time, though, the emergence of the United Kingdom as an oil-exporting nation has, particularly with regard to manufacturing,

122

brought out some of the employment implications of the so-called 'Dutch disease', which, coming just after the structural adjustments brought about by the UK's entry into the Common Market, appears to have reinforced any pre-existing independent UK 'de-industrialization' trends. The competing away of many traditional skill and craft-based labour-intensive manufacturing sectors raises the question to what extent, once oil has run out, and the economy has adjusted to its oil-depleted exchange rate, some of these skills can be regenerated relatively quickly and the lost shares of world markets be regained. Or will the exchange rate adjustment also mean a downward skill adjustment in the UK's international specialization pattern, forcing it to compete on the low wage end of the manufacturing spectrum with, one may fear, some severe balance of payments growth constraint?

We do not wish to add much to this debate here. The extent to which the pound's exchange rate is indeed oil 'dependent' remains very much open to debate and has, as the recent fall in oil prices has demonstrated, generally been overemphasized. The 'healing' from the effects of the Dutch disease will undoubtedly be a gradual process. It remains nevertheless a fair speculation that North Sea oil depletion will raise some of the severest challenges to sustained growth of the UK economy. This macroeconomic debate falls, however, outside the scope of the analysis in this chapter. We shall, return to it, albeit briefly, in part IV.

Our particular concern with manufacturing is based on a rather different distinctive feature of the manufacturing sector: that is, its particular role in both the generation and diffusion of new technologies. This predominant role relates as much to 'process-oriented' technologies, 'embodied' in new investment in manufacturing and/or services as to the more narrowly product-based innovations 'produced' (and used) in manufacturing but practically only used in services. This is, however, primarily an empirical argument and it will form the focus of most of the analysis in the first (technology) section of this chapter. It is also very much a time-bound argument. As we discuss in more detail in chapter 10, it could well be the case that in the near future more service activities will become fundamental in the generation and diffusion of new technologies. This is, as illustrated in chapter 10, already the case with regard to information technologies, e.g. in the 'service' component of the computer sector, software services are increasingly becoming the dominant technology-'producing' sector rather than the hardware manufacturing side. But as the subsequent analysis illustrates, even in this case close interaction with manufacturing (not just the computer sector, but also instruments, machinery, etc) is an absolute prerequisite for technological and economic success.

The crucial interdependence between manufacturing and a large number of service activities is undoubtedly insufficiently recognized. As pointed out in the next chapter on services, large parts of the tertiary sector are not just very similar to the manufacturing sector, in terms, e.g., of the capital-intensive nature of the services produced, but are also crucially manufacturing 'dependent'. As Momigliano and Siniscalco (1983) have shown, this implies that rather than a further tendency towards 'deindustrialization' in the 'postindustrial' Bell (1974) sense, the contraction of manufacturing over the last decade has rendered large parts of both manufacturing and service activities obsolete, bringing about structural change not just in the industrial part of the economy but also and increasingly in the service part.

The extent of these latter changes and their implications for future employment growth are discussed and illustrated in the next chapter. They also bring us to focus in a separate chapter on the emerging IT sector crossing the traditional manufacturing–service division.

The Generation and Diffusion of New Technologies

In chapter 4 we argued that contemporary technical change represents a change of techno-economic paradigm pervading all sectors. Our sector studies (Gower Series) confirmed the way in which computer-related innovations were penetrating manufacturing and service activities in the seventies and eighties. We attempt in this chapter first of all to relate this empirical evidence to the more systematic statistical data emerging from comprehensive national surveys of technological inputs and outputs.

The most traditional way of looking at a sector's effort in the generation and creation of new technologies consists of looking at how much a sector spends on research and development (R & D). Such an approach, focusing in the first instance on an input approximation of technological effort, ignores of course the differences between sectors in the efficiency of carrying out research and development and the extent of interindustry technology flows. It brings out, however, the over-riding importance of some sectors which appear to be at the origin of the generation of new technologies. It also (and in parenthesis) fits the production function economic approach, where alongside labour, capital and material inputs, R & D emerges as a crucial production factor input, opening up possibilities to estimate econometrically its marginal product as well as its contribution to economic growth (for such an attempt at the aggregate level, see Soete and Patel, 1985).

At a more pedestrian empirical level, R & D-intensive sectors appear to have been characterized by rapid output and productivity growth. Beginning our analysis with a closer look at the UK R & D expenditure distribution by sector is consequently probably the best starting point. Most recent UK figures for 1983 are given in table 8.1.

These 1983 figures are unfortunately based on a sample survey of only 75 private companies rather than on the normal full R & D survey of more than 700 companies, which was carried out in 1981 and is since then only carried out every 4 years. The next full survey will cover April 1985–April 1986, and will be published in 1987. Particularly the industry-product allocation of the R & D expenditure data might as a consequence have suffered from the heavy reliance on the estimation procedures used. The figures are nevertheless good enough to bring out the heavy concentration of R & D expenditure in a small number of sectors such as drugs, aerospace, telecommunications, computers and electronics, all with R & D-net output ratios above 15 per cent. In some of these R & D-intensive sectors the crucial importance of this 'intangible' investment is illustrated by the fact that the amount spent on R & D is well above the amount spent on 'physical' investment (as for example in the computer sector). The figures in table 8.1 point also to the very small amount spent on R & D in the non-manufacturing sectors, mainly in the so-called 'network' service sectors, such as electricity (CEGB), gas (BGC), railways (BR), etc. These sectors have on average seen their R & D expenditure increase less than in the manufacturing sector. On average and in real terms R & D expenditure in non-manufacturing has fallen. As we will see in chapter 10 the existing methods of identifying 'research and development' are however biased towards industrial production, in so far as they do not include, e.g. software expenditures.

But as we indicated before, R & D-expenditure data only offer one side of the picture which might in the final instance bear little resemblance to actual technological performance. Some alternative technology 'output' proxies exist. The number of patents filed or granted is a particularly useful proxy within an international context and has been used extensively (see e.g. Soete, 1980, 1981; Pavitt, 1980, 1983). At the intersectoral level, however, it raises too many questions about the difference in the propensity to patent between the various sectors, such as e.g. between aerospace and fabricated metal products, to be of much use here. We have, therefore, selected for closer examination an alternative output proxy based on innovation counts.

These data were gathered at the Science Policy Research Unit over the period 1970–85 and are based on a survey of some 4378 significant innovations introduced in the United Kingdom since 1945. Some use of the innovation data was already made on a sector-by-sector basis in the various Gower volumes. In the analysis presented here and in contrast to chapter 10, we will only use the aggregate sectoral data; no individual innovations will be listed or mentioned. The method of selecting and collecting these data has been explained in detail in Townsend et al. (1981), and Robson and Townsend (1984). Various analyses with regard to some of

Table 8.1 Intramural expenditure on R & D in 1981 and 1983

	Total R & D/VA £m 1981	Ratio	Total R & D/VA £m 1983	Ratio
Total[a]	3792.5		4163.3	
Extractive industries	62.6	1.67	76.1	2.22
Mineral oil refining	9.4	0.27	8.5	0.17
Other treatment of petrochemical products (excluding petrochemical manufacture)	27.8		22.4	
PRODUCTS OF MANUFACTURING INDUSTRY				
Total	3511.7		3869.9	
Iron and steel	31.7	2.02	26.9	1.57
Non-ferrous metals	21.6	3.60	23.5	2.86
Brick, cement, building materials etc.	12.9	1.06	14.1	0.94
Pottery, china and glass	20.2	1.43	21.5	1.28
Chemical Industry: total	617.4	11.54	735.0	11.08
Synthetic resins and plastics materials	42.4[d]	—[d]	40.0	—[d]
Paint	13.3	3.18	15.9	3.28
Pharmaceutical products	296.1	23.58	377.7	23.89
Other chemical products	265.7	8.38[d]	301.4	6.87[d]
Metal goods	22.7	0.71	2.5	0.63
Mechanical Engineering: total	234.0	2.87	249.6	2.96
Industrial plant and steelwork	15.0	1.14	18.0	1.23
Metal-working machine tools	13.0	1.97	13.2	2.14
Construction and earth-moving	34.2	3.09	34.4	3.15
Other machinery and equipment	171.8	3.39	184.0	3.50
Office machinery	14.1	24.03	10.7	25.16
Electronic data processing equipment	160.8		247.3	

...d electronic engineering, total	1101.1	11.85[a]	1535.6	16.75[c]
Insulated wires and cables	16.3	4.44	11.5	2.66
Basic electrical equipment	55.6	5.45	58.8	4.73
Telegraph and telephone apparatus	279.9	18.87[e]	360.3	20.45[e]
Electrical instruments and control systems	67.9	15.40	72.3	13.02
Radio and electronic capital goods	610.4	57.98	633.3	51.16
Components other than active	26.3	6.62	49.7	10.45
Active components and sub-assemblies	55.2	18.36	77.6	21.26
Electronic consumer goods	20.8	8.32	22.7	7.37
Other electrical goods	48.8	4.08	47.4	3.62
Motor vehicles and parts	180.4	5.62	239.5	6.69
Shipbuilding and repairs	9.5	0.89	8.4	0.90
Aerospace equipment manufacturing and repairing	762.9	37.57	720.0	28.94
Instrument engineering	60.6	7.85	49.4	5.32
Food, drink and tobacco	91.5	1.09	80.2	0.84
Textiles other than man-made fibres	9.7	0.54	10.2	0.52
Leather, footwear and clothing	4.7	0.23	5.3	0.24
Timber and wooden furniture	3.6	0.20	3.9	0.19
Paper and paper products: printing and publishing	18.3	0.34	21.4	0.35
Processing of rubber and plastics	30.0	1.43	24.7	1.02
Other manufactured products	24.1	2.56	22.0	2.22
CONSTRUCTION[b]	15.3	0.16	15.7	0.14
UTILITIES AND SERVICES[c]	165.6	2.36[f]	170.8	2.11[f]

[a] Analysis in the table uses the SIC (1980).
[b] The figures for this product group do not include estimates for the appreciable development work undertaken in civil engineering.
[c] Gas, electricity, water, railway, road and sea transport systems, industrial health and safety systems, etc.
[d] Synthetic resins included in other chemical products.
[e] Excluding BT.
[f] Only utilities.
Source: Calculated from *British Business*, 18 January 1985 and *Business Monitor*.

Table 8.2 Technology classification of innovative industries. (>20 innovations produced or used between 1945 and 1983.)

Sector (MLH)	Nr. innov. produced	Nr. sectors used	Own use	Manuf. use	Service use	Main user sector	%	Nr. innov. used	Main origin sector	%
A INNOVATION PRODUCING SECTORS (Production > use)										
1 Pervasive sectors (≥20 sectors UK)										
Scientific & industrial instruments (354)	424	61	36	278	110	Textile fibres	11.6	93	Electronic components	22.6
Other machinery (339)	393	62	17	331	45	Coal mining	30.3	44	Metal ind. n.e.c.	22.7
Metal-working machine tools (incl. robotics) (332)	191	20	10	179	2	Mechanical eng.	56.5	23	Instruments	47.8
Electronic capital goods (367)	188	31	16	35	137	R & D services	18.1	77	Electronic components	62.3
Electronic components (364)	183	26	41	118	24	Electronic cap. goods	26.2	55	Glass	5.5
Electronic computers (366)	106	28	25	26	55	Other business services	11.3	47	Electronic components	23.4
Industrial plant & steelworks (341)	97	29	29	57	11	Steel	23.7	41	Steel	7.3
Plastics, n.e.c. (incl. composite materials) (496)	83	38	8	54	21	Construction	19.3	25	Other machinery	28.0
Synthetic resins, plastics (276)	79	27	8	57	14	Motor vehicles	15.2	20	Other machinery	25.0
Pumps, valves and compressors (mechanical ... (333)	7?	26	22	41	7	Construction	10.0	28	Instruments	7.1

2 *Localized sectors* (≤ 10 sectors use)

(a) *Localized sectors*, with predominant own use of *own* produced innovations

Sector									
Radio & TV equipment (365)	74	9	40	34	Other services	20.3	61	Electronic components	19.7
Tractors (380)	48	3	26	22	Agriculture	39.6			
Locomotives & railways (384)	37	7	9	28	Railways	40.5	23	Instruments	47.8
Soap & detergents (275)	26	5	17	9	Other services	23.1			
Cement (464)	26	2	14	12	Construction	46.1			
Motor cycles (382)	22	2	17	5	Other services	22.7	20	Other machinery	25.0

(b) *Localized user-dependent sectors*

Sector						
Textile machinery (335)	232	3	3	229	Textile fibres	98.3
Drugs (272)	76	4	1	75	Medical and dental services	96.1
Dyestuffs & pigments (277)	48	8	8	40	Textile fibres	62.5
Agricultural machinery (331)	39	5	6	33	Agriculture	76.9
Surgical instruments (353)	28	5	4	24	Medical and dental services	75.0
Construction machinery (336)	48	10	10	38	Construction	50.0

(continued)

Table 8.2 *(continued)*

3 Users-influenced

Sector (MLH)	Nr. innov. produced	Nr. sectors used	Own use	Other use (Manuf. & Services)	Main user sector	%	Nr. innov. used	Main origin sector	%
Shipbuilding (370)	134	14	40	94	Petroleum & natural gas	38.1	111	Metal working machine tools	17.0
Electrical machinery (361)	68	16	17	51	Electricity	38.2	30		
Photographic equipment (351)	66	13	9	57	Other services	30.3			
Glass (463)	57	16	14	43	Construction	15.8	32	Instruments	32.0
Other chemical industries (279)	54	13	2	52	Agriculture	35.2			
Non-ferrous metals (321)	53	12	29	24	Mechanical engineering	13.2	33		
Other electrical goods (369)	53	12	6	47	Motor vehicles	37.7			
Insulated wires & cables (362)	46	11	19	27	Postal services & telecommunications	26.1	32		
General chemicals (271)	42	15	17	25	Paper & board	14.3	37		
Telecommunications (363)	37	14	3	34	Other business services	37.8	34		
Mechanical handling (337)	31	16	3	28	Construction	35.5			

B INNOVATION-USING SECTORS (Nr. using > Nr. producing)

1 With significant own technological potential

Sector (MLH)	Nr. innov. produced	Nr. sectors used	Own use	Other use (Manuf. & Services)	Main user sector	%	Nr. innov. used	Main origin sector	%
Aerospace (???)	135	8	99	36	National government service		172	Instruments	9.9

Sector									
Motor vehicles (381)	111	12	82	29	Road passenger transport	185	Other electrical machinery		10.8
Ferrous metals (311)	102	18	54	48	Shipbuilding	120	Instruments		20.8
Food (211)	70	8	37	33	Wholesale distribution	79	Instruments	30.0	24.0
Construction (500)	41	10	22	19		182	Construction machinery		13.2
Paper (481)	36	15	13	23	Electronic components	55	Other machinery		47.3
Leather (431)	35	4	28	7		44	Other machinery		13.6
2 Supplier-dependent industrial sectors (one supplier > 50% of Nr. in use)									
Mechanical engineering (330)	1	0	0	1		172	Metal working machine tools		62.8
Man-made fibres (textiles) (411)	96	8	62	66	Unknown	387	Textile machinery		58.9
Coal-mining (101)	8	8	8	0		190	Other machinery		62.6
(Other) printing machinery (489)	5	3	3	2		82	Other machinery		61.0
Petroleum & natural gas (104)	4	4	4	0		81	Shipbuilding		63.0
3 Supplier-dependent services (one supplier > 50% of total Nr. in use)									
Electricity (602)	1	0	0	1		40	Electrical machinery		65.0
Wholesale distribution (810)	0	0	0	1		31	Food		67.7
Postal & Telecommunications services (708)	5	0	0	5		23	Insulated wires		52.2
Cinemas, Theatres (881)	0	0	0	0		22	Radio, TV		50.0

(continued)

Table 8.2 (continued)

Sector (MLH)	Nr. innov. produced	Nr. sectors used	Own use	Other use (Manuf. & Services)	Nr. innov. used	Main origin sector	%
4 Suppliers-influenced							
Medical services (874)	4		4	0	166	Drugs	44.0
National government services (901)	0		0	0	136	Electronic capital goods	30.2
Other (Unknown) services (950)	8		0	8	119	Photographic equipment	16.8
Research & development services (876)	3		0	3	117	Instruments	39.3
Agriculture (001)	6		2	4	93	Agricultural machinery	32.3
Other business services (865)	2		1	1	53	Telecommunications	26.4
Sea transport (705)	2		2	0	43	Shipbuilding	37.2
Local government services (906)	8		0	8	37	Electronic capital goods	21.6
Gas (601)	16	1	16	0	35	Instruments	14.3
Air transport (707)	0		0	0	29	Electronic capital goods	41.4
Railways (701)	0		0	0	28	Railway equipment	43.4

the main features of the data have been published by Pavitt (1983, 1984) and Pavitt, Robson and Townsend (1985a, b). Our analysis will closely follow Pavitt's attempt to develop some sort of sectoral taxonomy of technical change using these innovation counts data. In contrast to his richer analysis, we will limit ourselves to the sector-based data, leaving aside the company-based data.

One of the most interesting features about the innovation survey data is that they also contain data by sector of use and not just by sector of origin. This actually might help one to overcome one of the major drawbacks of technology proxy data mentioned earlier on: that is, their exclusive concern with the generation, i.e. the 'production' of new technologies. This means that the data can be used virtually as a kind of 'input–output' table of innovations. Some sectors might well be innovative in their 'first' use of an innovation, rather than in its generation. Table 8.2 represents a relatively straightforward grouping of all 'innovative' sectors, i.e. sectors which either as 'producers' or as 'first users' had accounted for more than 20 innovations in the United Kingdom over the period 1945–83. These various sectors account for some 96 per cent of the total number of innovations used.

The various MLH sectors (the data were gathered by MLH rather than Activity Headings) were grouped according to three different criteria:

1 Was the sector predominantly innovation-producer or innovation-user?
2 Were the innovations originating from one sector, pervasive in their user impact, i.e. were they used in many other sectors (more than 20), or on the contrary localized in their user impact, i.e. used only in a few (less than 10) sectors?
3 Did any predominantly innovation-producer sector rely heavily on one user in the use of its innovations, i.e. was it 'user-dependent' (more than 50 per cent) or was it itself its own main user, i.e. 'user-independent'; and by the same token, did any predominantly innovation-using sector rely heavily on one supplier in the production of its innovations, i.e. was it 'supplier-dependent' (more than 50 per cent) or was it itself its own main producer, i.e. 'supplier-independent'?

Combining these three criteria, the following seven different groups of 'innovative' sectors could be identified.

Pervasive sectors

As illustrated in table 8.2, part A1, the MLH sectors included under this heading correspond in many ways to what one might have expected a priori. They represent indeed some of the most influential and pervasive technologies, such as instruments (MLH 354), electronics (367 and 364), computers (366), plastics (496 and 276) and robotics (332). Interestingly, though one also finds under this heading process plant engineering (341), one of the most innovation-embodied, technology-supplying industries, and two more heterogeneous sectors: other machinery (339), which practically by definition would see its use spread over a large number of using sectors (coal-mining, food-processing, wood, furniture, paper, printing, etc); and pumps, valves and compressors (333), which in some ways could be described as 'mechanical instruments'.

Of the various sectors listed under A1, it is worth observing that only the computer and the electronic capital goods sector are also pervasive with respect to widespread first use in the service sector. All other sectors are only pervasive with regard to manufacturing.

The absence of some highly R & D-intensive sectors, such as aerospace and drugs, is another feature worth noting. These sectors were part of the so-called 'science-based' sector in Pavitt's taxonomy of technical change. Their absence here highlights the fact that R & D-intensity cannot be equated with 'pervasiveness'.

Localized Sectors

The sectors listed in table 8.2, part A2, are sectors which have been highly innovative in their own right, but have not made much 'user-impact' on many sectors. A first group listed under A2a consists of sectors which do not depend on one single sector in the use of their innovations. The most innovative one included under this heading is radio and TV equipment (365). Most sectors are also highly innovation-using sectors.

This is a typical characteristic of the sectors grouped under this heading; whereas they are significant innovation-producing sectors, they rely heavily on innovations generated elsewhere and their impact as innovation-supplier on the rest of the economy remains limited.

User-dependent Sectors

A special group of localized, predominantly innovation-producing sectors (grouped in table 8.2, part A2b, are also user-dependent, i.e. more than half of their innovations are used by one single sector. Textile machinery (MLH 335), for example, is dependent on textile fibres in its use of innovations; drugs (272) on medical and dental services; dyestuffs and pigments (277) on textile fibres; agricultural machinery (331) on agriculture; surgical instruments (353) on medical and dental services; and so on.

In contrast to the previous group of sectors, the sectors under this heading are not just limited in their impact on other sectors, they are also practically only innovation-producing sectors; their first use of innovations is minimal. They could be compared to islands of innovative activity in the economy. This does not mean that some of the sectors listed here, such as drugs, would not be amongst the most R & D-intensive sectors of the economy.

User-influenced Sectors

The remainder of the predominantly innovation-producing sectors are neither dependent on one specific using sector, nor really pervasive in their impact on other sectors. They are listed in table 8.2, A3. Most innovative here are electrical machinery (MLH 361) with first use spread over a number of service (electricity generation) and manufacturing sectors; photographic equipment (351); glass (463); other chemical products (279); non-ferrous metals (321); other electrical goods (369); general chemicals (271); telecommunications (363) and mechanical handling equipment (337).

Some of these sectors, such as glass, electrical machinery or mechanical handling equipment, have a relatively widespread user impact. None, however, not even telecommunications, is 'pervasive' in the spread of its user impact. In interpreting these data one should remember that the innovations surveyed relate to the overall postwar period. Particularly for the most recent period, the number of significant innovations has probably been underestimated.

Predominantly Innovation-using Sectors, with a Significant Own Technological Potential

Some of the sectors grouped in table 8.2, B1 are only marginally more innovation 'using' than 'producing'. All sectors have a significant own technological potential. The list includes aerospace, motor vehicles, iron and steel and food as the most innovative ones. Nevertheless, they are all more innovative in their use of others' innovations than in their own innovation generation potential. However, they are also characterized by the high use of their own 'produced' innovations.

Rather than pervasive in their impact, these sectors, it could be argued, are 'absorptive' in their use and adaptation of innovations generated elsewhere. They are however, as in the case of the 'pervasive' sectors reviewed under A1, at the centre of the process of technological change, not so much in the actual generation of innovations as in their widespread use and adaptation.

This might explain why some of the sectors listed under this heading are indeed relatively R & D-intensive sectors, such as aerospace, or motor vehicles, the large amounts of R & D outlays relating, apart from the sector's own produced and used innovations (in Pavitt's terminology the sector's process innovations), to the necessary adaptation to the sector's requirements of a large amount of innovations generated elsewhere and first used in the sector.

Supplier-dependent Industrial Sectors

With the exception of textile fibres, the sectors included under this heading and listed in table 8.2, part B2, are not only dominated by a single supplying sector in the number of innovations they first use, they also appear to have very little technological potential of their own.

Textiles, by contrast, is dominated by one supplier in its innovation use, but has, particularly with regard to the textiles fibres subfield, a significant own innovation production potential.

Supplier-dependent/Suppliers-influenced Services Sectors

A number of services also appear to rely heavily on a single innovation supplier. The service sectors listed in table 8.2, part B3, as in the next group in part B4, do have practically no innovation 'production' potential. They are all innovative only with regard to the large number of first use of innovations generated and produced elsewhere. This heavy reliance on 'use' of innovations rather than 'production' explains why the service sector will be so under-represented with respect to traditional technology input or output proxies. The sectors listed in B3 are supplier-dependent with regard to this innovation-user potential; the sectors listed in B4 by contrast do not rely on a single innovation-producing sector.

The technological classification attempted above and detailed in table 8.2 has a number of shortcomings, directly related to the (sometimes questionable) way in which innovations have been allocated to producers' and users' MLH sectors. Some MLH sectors are clearly under-represented, others (particularly those ending with 1 or 9) are significantly over-represented. The grouping of sectors attempted here is only as good as the data.

However, it brings out a number of broad features which systematically support some of the arguments put forward earlier on with regard to the importance of manufacturing in the generation and production of new technologies and the claim made in chapter 4 as to the 'pervasiveness' of some of the new information technologies such as microelectronics and computers and such industries as instrumentation, which has been virtually absorbed by microelectronics.

They also point to the importance of some other technologies, such as plastics/composite materials as having a 'pervasive' impact on a widespread number of other sectors. However, in contrast to the electronics and computer technologies, the user impact of these other technologies appears to be limited to manufacturing sectors. The all-round pervasiveness of the cluster of IT technologies is addressed at greater length in chapter 10. We now turn to some of the broad economic trends in the various manufacturing sectors.

Sectoral Trends in Manufacturing Growth

The purpose of this economic section is to provide a brief overview of some of the trends in employment, output, investment and productivity growth in the various UK manufacturing sectors. We start the analysis with a look at the broad postwar trends in employment in a number of manufacturing sectors. As illustrated in figure 8.1, the analysis, in contrast to the sectoral studies published in the Gower Series, is in terms of the 1980 SIC classes rather than in terms of the 1968 MLH classes. Economic information by MLH class is no longer updated by the CSO or the DoE. While the 1980 SIC is more closely comparable with the EEC NACE classification system, and therefore allows more easily for international comparisons, it is unfortunately a rather poor classification system; for example, it actually no longer allows for a clear-cut separation between the electrical and electronic sectors at the 3-digit level, something which both from a technological and economic perspective is essential.

In addition to the sectors separated out in figure 8.1, based on the highest level of disaggregation available from the CSO with regard to capital stock estimates we have therefore also estimated, following the method developed in Soete (1985a), an electronic sector consisting of Activity Headings 3443 (radio and electronic capital goods); 3444 (non-active electronic components); 3452 (active electronic components) and 3553, 3554 (electronic consumer goods). For the period before 1981, the available data on the corresponding MLH sectors (367, 364, and 365) have been used as calculated in Soete and Dosi (1983). Electrical, including telecommunications (AH: 3441 or MLH 363), is then identified as the remaining 'non-electronic' component of SIC Class 34: electrical and electronic engineering. Electronic computers or data processing equipment becomes, however, a separate 2-digit Class (1980 SIC: 33) and is based on the official CSO data. All data used cover the period 1948–84. Five subperiods will be considered: 1948–55; 1955–63; 1963–73; 1973–9 and 1979–84 (with the exception of 1948 and 1984, all peak years).

Figure 8.1 illustrates the dramatic decline in employment in manufacturing over the postwar period. Only the electronics, computer, food and electrical industries witnessed any significant employment growth over the full period 1948–84. For the most recent period, 1979–84, all manufacturing sectors except the computer sector saw their employment fall. The annual detail of the sectoral data used can be found in the appendix to this chapter.

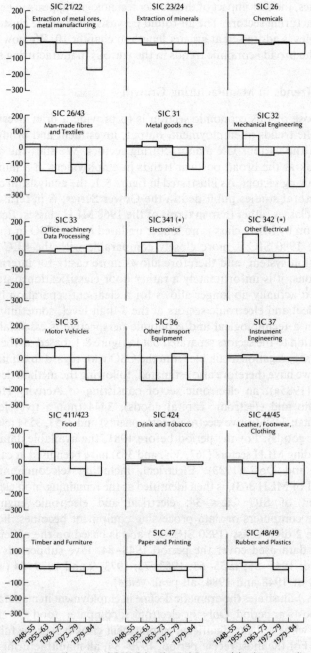

(*) SIC 341 Electronics: SIC 342 Other Electrical. For annual detail see appendix

Figure 8.1 Employment trends (1948–84) for 5 periods

Table 8.3 Employment growth* (decline −) in the manufacturing sectors (GB by activity heading, 1981–86)

Sector	AH	Employment
Radio and electronic capital goods	3443	8,600
Measuring, precision instruments etc.	371	7,400
Builders' carpentry and joinery	463	4,100
Other electronic equipment	345	3,600
Components other than active components	3444	3,300
Office machinery, data processing equipment	33	2,900
Pharmaceutical products	257	1,200
Medical and surgical equipment	372	0,800
Electric lighting equipment and electrical installation	347/348	0,400
Refrigerating machinery, space heating, ventilation	3284	0,200
Insulated wires and cables	341	−0,200
Non-ferrous metal foundries	3112	−0,200
Saw-milling, planing, semi-finished wood products	461/462	−0,300
Shop and office fitting	4672	−0,400
Specialized household products	259	−0,400
Compressors and fluid power equipment	3283	−0,500
Printing and publishing of books etc.	4752/4753	−0,600
Building products of concrete, cement etc.	243	−0,700
Work clothing and men's and boys' jeans	4534	−0,900
Optical precision instruments etc.	373	−0,900
Photo/cinematographic processing	493	−1,000
Domestic-type electric appliances	346	−1,500
Telecommunication equipment	344	−1,500
Structural clay	241	−1,600
Machinery for food etc. industries	324	−1,800
Textile machinery	323	−1,800
Soft drinks	428	−2,300
Milk and milk products	413	−2,500
Machinery for printing etc. industries	327	−2,500
Metal doors, windows etc.	314	−2,500
Soap and toilet preparations	258	−2,500
Industrial equipment, batteries etc.	343	−2,600
Asbestos goods	244	−2,600
Fish processing	415	−2,700
Other manufacturing nes	492–495	−2,800
Articles of wood, cork etc.	464/465/466	−2,800

(continued)

Table 8.3 *(continued)*

Sector	AH	Employment
Bolts, nuts, springs etc.	313	−2,900
Man-made fibres	26	−2,900
Ordnance, small arms and ammunition	329	−3,000
Sugar and sugar by-products	420	−3,100
Fruit and vegetable processing	414	−3,200
Paints, varnishes and printing ink	255	−3,200
Men's and boys' tailored outerwear	4532	−3,300
Cycles, motor cycles and other vehicles	363–365	−3,300
Jewellery and coins	491	−3,500
Cement, lime and plaster	242	−3,600
Abrasive products and working of stone etc.	245/246	−3,700
Household textiles etc.	455	−3,800
Bacon curing and meat processing	4122	−3,800
Leather and leather goods	44	−4,000
Meat and meat products, organic oils and fats	411/412	−4,000
Engineers' small tools	3222	−4,100
Hand tools and finished metal goods	316	−4,100
Packaging, production of board	4725	−4,200
Clocks, watches etc.	374	−4,200
Steel drawing, cold rolling, cold forming	223	−4,400
Carpets etc.	438	−4,600
Textile finishing	437	−4,700
Mechanical lifting and handling equipment	3255	−4,900
Pulp, paper and board	471	−5,300
Wooden upholstered furniture	4671	−5,300
Copper, brass and other copper alloys	2246	−5,300
Spirit distilling and compounding	424	−5,500
Refractory and ceramic goods	248	−5,500
Printing and publishing of newspapers	4751	−5,600
Specialized industrial products	256	−5,600
Aluminium and aluminium alloys	2245	−5,600
Other textiles	433/434/435/439	−5,700
Forging, pressing and stamping	312	−5,800
Toys and sports goods	494	−6,000
Ferrous metal foundries	3111	−6,000
Processing of plastics	483	−6,300
Cotton and silk	432	−6,400
Agricultural machinery and tractors	321	−6,500
Metal-working machine tools	3221	−6,800
Cocoa, chocolate, sugar confectionery etc.	421	−6,900
Women's and girls' tailored outerwear	4533	−7,200

Table 8.3 *(continued)*

Sector	AH	Employment
Printing and publishing	475	− 7,300
Basic electrical equipment	342	− 7,500
Steel tubes	222	− 7,500
Animal feeding stuffs and miscellaneous food	416/418/422/423	− 7,600
Inorganic chemicals except inds gases	2511	− 7,600
Mechanical power transmission equipment	326	− 7,700
Footwear	451	− 8,100
Conversion of paper and board	472	− 8,400
Women's and girls' light outerwear, lingerie etc.	4536	− 8,900
Hosiery and other knitted goods	436	− 9,400
Glass and glassware	247	− 10,300
Bread, biscuits and flour confectionery	419	− 10,400
Brewing and malting, cider and perry	426/427	− 10,500
Bodies, trailers and caravans	352	− 10,800
Woollen and worsted	431	− 11,700
Tobacco	429	− 11,900
Industrial plant and steelwork	320	− 11,900
Mining machinery etc.	325	− 13,800
Railway and tramway vehicles	362	− 13,900
Telegraph and telephone apparatus and equipment	3441	− 14,400
Internal combustion engine except road vehicles etc.	3281	− 16,000
Non-ferrous metals	224	− 16,200
Motor vehicles and engines	351	− 16,700
Basic industrial chemicals	251	− 17,300
Rubber products, tyre repair etc.	481/482	− 17,600
Clothing, hats, gloves and fur goods	453–456	− 25,900
Aerospace equipment	364	− 27,500
Other machinery and mechanical equipment	328	− 29,900
Iron and steel	221	− 30,800
Parts	353	− 31,100
Shipbuilding and repairing	361	− 33,900

* The period covers December 1981 – March 1986
Source: Employment Gazette, August 1984 and August 1986

The level of disaggregation used remains, however, poor. Unfortunately, more disaggregated employment data (at the Activity Heading level) are only available from 1981 onwards. These, as table 8.3 illustrates, might nevertheless provide a more detailed insight into the extent of the employment growth and decline at the detailed 3- and 4-digit AH level over the most recent period. At least, one advantage of the lack of comparable sectoral data before 1981, is that the choice of the period analysed (December 1981–March 1986) is out of one's control, allowing us to add a possibly more neutral picture to the debate about the UK's success in employment growth over recent times.

The decline in overall employment over the period 1981–86 is clear-cut (– 195,000). The growth in service employment (+ 675,000), particularly distribution (+ 246,700) and banking, insurance and business services (+ 285,300), has been insufficient to offset the decline in employment in the industrial (– 837,200), transport (– 108,700) and government (– 31,300) sectors. Within manufacturing, the issue of more direct concern to us here, the decline in employment (– 586,500) is widespread and affects practically all sectors. The exceptions are limited to: AH 3443, electronic capital goods (+ 8600); AH 345, other electronic equipment (+ 3600); AH 371, measuring instruments (+ 7400); AH 3444, electronic components (+ 3300); AH 463, builders' carpentry and joinery (+ 4100); AH 330, data processing equipment (+ 2900); AH 257, pharmaceutical products (+ 1200); AH 372, medical instruments (+ 800); AH 3470, 3480, electrical lighting equipment (+ 400) and AH 3284, refrigerating machinery, space heating, ventilation (+ 200). The predominance of electronic and instrument sectors in this list is striking, and fits our own employment forecasts for the electronic sector (Soete and Dosi, 1983; Soete, 1985a). It also suggests that while at first glance there would appear to exist some relationship between technology (R & D)-intensity and employment growth (only some though: in contrast to electronics, computers and drugs, employment fell by some 27,000 in the aircraft industry, and by some 28,000 in the chemical industry) a closer relationship appears between employment growth and the pervasiveness of the sector. On the other side of the spectrum, there appears little improvement in the continuous heavy loss of employment in the mechanical engineering, metal manufacturing, motor vehicles and other transport industries.

These sectors have replaced textiles, clothing and leather goods in the sheer size of the job displacement which has occurred. In parenthesis it is worth noting that this employment decline in the vehicle manufacturing industries (and their main suppliers) goes hand-in-hand with a significant employment decline in the transport service sector, pointing again to the close interaction between manufacturing and services. The same argument holds for computer services, where the growth in the manufacturing computer sector goes hand-in-hand with a significant growth in AH 8394, computer services (+ 11,300). With regard to many other 'business' services which have witnessed significant employment growth, it is clear though that more straightforward substitution between manufacturing and services through increased subcontracting of particular service activities has occurred. The increases in AH 923, cleaning services (+ 29,900) and AH 84, renting of removables (+ 11,700) are indicative of this trend. We turn to some of these issues in greater depth in the next chapter and to the specific employment growth potential of the IT-sectors in chapter 10.

Compared to the sectoral analyses published in the Gower volumes the use of the 1980 SIC will allow for some major improvements with regard to trends in output. The output data used here relate to *net* output rather than gross output. This should provide us with a more reliable and correct approximation of the actual output and productivity trends than estimates based on gross output. All data used cover the period 1948–84. Five subperiods will be considered: 1948–55; 1955–63; 1963–73; 1973–9 and 1979–84.

Output growth rates in real terms for the various subperiods for each sector are given in figure 8.2. The underlying annual data are given in the Appendix. The decline in output growth over the whole period is most dramatic in the motor vehicles manufacturing sector. Net output in 1984 was at about 70 per cent of its 1973 values. However, the recession seems to have affected most industrial sectors. Only in the computer and electronics sectors was output growth for the period 1979–84 higher than for the previous period. The evidence of widespread economic recovery in manufacturing based on the resumption of economic growth over the last couple of years (1983–5) remains consequently, within its broader historical perspective, rather limited. Output levels in most sectors were even in 1985 still well below 1979. Only in computers, electronics, chemicals and food did real economic resurgence occur and output levels are today (1985) at their historical height.

Investment outlays typically fluctuate quite heavily and within this context both growth rates and annual changes are rather meaningless. Figure 8.3 represents therefore the average *levels* of investments (in 1980 prices) for each of the five subperiods considered so far, for the various manufacturing sectors. Investment outlays in the ferrous and non-ferrous metal goods and mechanical engineering sectors as well as in the chemicals and textiles sectors have fallen significantly over the most recent period. Typically the very high investment outlays in these sectors over the period 1973–9 with the onslaught of the recession came at the worst moment. The decline in output over this period was to a large extent unexpected, leading to significant overcapacity in most sectors.

The lower, but still substantial, recent investment outlays in these sectors are in all likelihood different in nature from those of the previous period and can be expected to be more directed towards restructuring and rationalization, aiming at reducing labour costs and resulting, as figure 8.1 illustrates, in significant employment displacement.

A similar trend probably also underlies those sectors, such as motor vehicles, other transport equipment, food, drink and tobacco, and paper

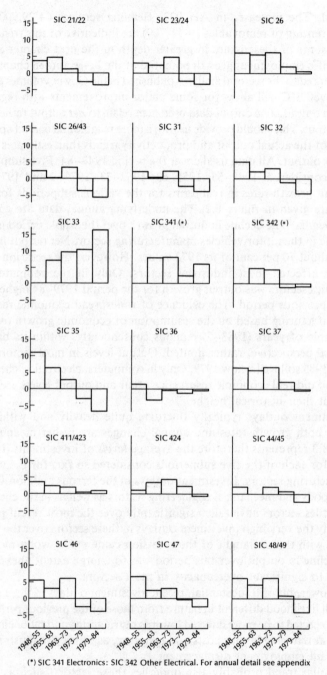

(*) SIC 341 Electronics: SIC 342 Other Electrical. For annual detail see appendix

Figure 8.2 Output growth by sector (1948–84 for 5 periods)

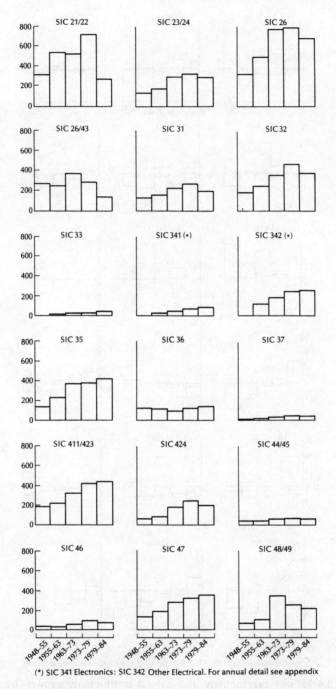

(*) SIC 341 Electronics: SIC 342 Other Electrical. For annual detail see appendix

Figure 8.3 Average investment outlays by sector (1948-84 for 5 periods)

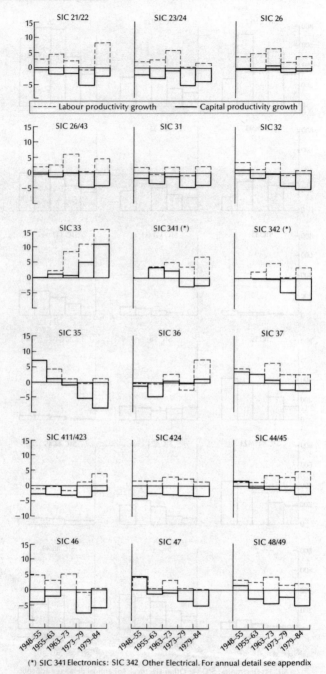

(*) SIC 341 Electronics: SIC 342 Other Electrical. For annual detail see appendix

Figure 8.4 Labour productivity and capital productivity growth by sector (1948–84 for 5 periods)

and printing, which saw their investment outlays increase over the most recent period despite very low, if not negative, output growth (see figure 8.2).

It is only in the computer industry, and to a lesser extent the electronics industry that one finds clear evidence of a 'virtuous' growth pattern of high and increasing investment outlays, rapid output growth and, in the case of the computer sector, employment creation.

The extent to which investment outlays were of the expansionary (capital-widening) or rationalization (capital-deepening) type will to some extent be reflected in the trends in labour productivity growth. The high investment outlays in the transport and paper and printing sectors, accompanied by falling output levels, will also have led to a further increase in the marginal capital–output ratio. Only if this investment boom has been accompanied by the scrapping of large parts of existing obsolete capital stock will 'average' capital productivity have increased. It is to trends in these sectoral labour and capital productivities that we turn now.

Support for an increased trend towards rationalization investment emerges from figure 8.4 which represents the growth rates in labour productivity and capital productivity for the various manufacturing sectors, again for each of the five subperiods considered so far.

Labour productivity growth over the most recent period (1979–84), with the exceptions of drink and tobacco, and paper and printing, has recovered in all sectors from its dramatic decline over the period 1973–9. In the cases of metal manufacturing (iron and steel), and other transport equipment, labour productivity growth is today at all time record levels. In these two sectors, capital productivity too has recovered substantially. In the chemical and textiles (including textile fibres) sectors, labour productivity growth is also at levels well above the average for the whole period. The recovery in labour productivity over the most recent period (1979–84) is indicative of the fundamental restructuring and rationalization pattern which has occurred in British manufacturing industry in recent times. This trend has been accompanied by massive employment displacement, and up to now a decline in output in many sectors.

In terms of 'capital' productivity, figure 8.4 illustrates the low, more often than not negative, growth in the output–capital ratio. The very negative growth rates over the period 1973–9 suggest that the record levels of investment expenditure over this period have indeed led to significant overcapacity in many sectors, particularly in ferrous and non-ferrous metals, textile fibres, metal goods, mechanical engineering and timber and wood. The (positive) growth in capital productivity over the most recent period in ferrous metals, motor vehicles and other transport industries points to an increased, more efficient use of capital in these sectors as well as accelerated scrapping.

148   P. Patel and L. Soete

The major feature emerging from figure 8.4, however, relates to the positive growth in capital 'productivity' over the period 1948–1984, in our three most 'pervasive' sectors: computers, electronics and instruments. These sectors emerge as the only ones with a simultaneous growth in both labour and capital productivity. Their importance, at the core of the new emerging IT sector, is discussed in greater detail in chapter 10.

Conclusion

This chapter has concentrated on two main issues in UK manufacturing: the ways in which technical change has affected employment over the postwar period, and the embodiment of that technical change in new investment. It has tried to show that, so far from causing the big decline in manufacturing employment, microelectronic-based sectors have been the main source of what little employment growth has occurred in the last few years in manufacturing. The decline of employment in the other sectors of UK manufacturing has, however, been massive. It was primarily the result of a process of restructuring, rationalization and plant closures in response to international competition pressures and depressed levels of demand for 'older products'. It was mainly 'deepening' rather than broadening the capital stock and represented the exhaustion of the growth potential of an older technological regime, based on mass production and dedicated capital equipment.

We now take up the issue of trends in service employment and how those relate to the trends in manufacturing.

Appendix Chapter 8

Sectoral Data for the UK Used in Chapter 8

Net Output (1980 £, millions), Employment ('000's), Investment (1980 £, millions) and Capital Stock (1980 £, millions) by SIC Sector (1948-84)[a]

SIC 21 and 22 Extraction of Metalliferous Ores, Metal Manufacturing

Year	Net Output	Employment	Investment	Capital Stock
1948	1943.03	468.400	221.30	4300.30
1949	2021.17	470.400	275.50	4477.10
1950	2109.73	475.400	313.30	4689.80
1951	2198.28	481.600	329.10	4915.70
1952	2286.84	488.500	291.80	5103.60
1953	2242.56	480.000	292.00	5287.00
1954	2388.42	481.200	383.40	5558.30
1955	2526.46	496.700	376.50	5820.60
1956	2544.69	504.100	437.00	6143.50
1957	2635.86	503.800	533.60	6563.70
1958	2500.42	488.000	542.40	6992.60
1959	2552.51	483.200	500.00	7377.70
1960	2776.50	520.400	733.00	7995.20
1961	3010.92	533.700	987.20	8859.20
1962	2995.29	502.200	832.80	9564.30
1963	3641.23	498.900	513.50	9943.30
1964	4138.71	522.900	418.70	10212.40
1965	4323.64	532.000	391.30	10440.70
1966	4081.41	520.400	382.00	10643.40
1967	3841.79	495.600	451.00	10913.60

(continued)

149

Appendix *(continued)*

Year	Net Output	Employment	Investment	Capital Stock
1968	4089.22	485.600	397.00	11123.80
1969	4185.59	488.100	423.40	11354.50
1970	4078.80	493.100	632.60	11794.30
1971	3641.23	463.100	782.20	12381.30
1972	3638.63	436.100	735.50	12916.00
1973	4039.73	436.400	622.70	13322.20
1974	3706.35	430.800	762.80	13848.90
1975	3203.66	427.400	883.80	14477.10
1976	3422.44	403.600	969.00	15169.30
1977	3370.35	415.900	788.70	15664.00
1978	3302.63	392.500	538.70	15905.20
1979	3440.68	383.600	428.10	16031.10
1980	2604.60	349.500	343.00	16059.60
1981	2766.10	269.000	238.00	15969.90
1982	2703.60	253.000	202.10	15816.60
1983	2714.00	225.100	183.00	15568.00
1984	2862.40	210.000	229.00	15285.10

SIC 23 and 24 Extraction of Mineral Ores and Non-Metallic Mineral Products

Year	Net Output	Employment	Investment	Capital Stock
1948	2077.51	374.600	137.40	1358.80
1949	2170.29	380.400	123.50	1447.00
1950	2299.38	388.700	122.10	1532.30
1951	2448.64	395.800	120.40	1616.00
1952	2396.20	399.000	105.10	1683.40
1953	2444.60	394.900	129.00	1774.60
1954	2493.01	397.100	136.60	1871.70
1955	2589.83	401.400	139.20	1970.90
1956	2561.59	396.700	152.30	2083.50
1957	2480.91	389.100	159.20	2202.60
1958	2440.57	376.200	149.30	2315.00
1959	2605.96	379.400	140.90	2419.40
1960	2872.21	390.300	194.60	2577.90
1961	3021.47	398.000	232.00	2773.20
1962	3057.77	401.300	243.30	2977.70
1963	3158.62	389.300	201.50	3138.80
1964	3662.87	402.400	220.80	3314.50
1965	3711.28	405.600	280.40	3547.60
1966	3715.31	398.200	291.90	3788.40
1967	3864.57	381.100	257.90	3991.40
1968	4134.85	384.700	354.40	4277.30
1969	4259.90	384.300	365.60	4576.80
1970	4150.99	375.100	313.20	4820.10
1971	4425.30	359.200	311.00	5058.00

Appendix (continued)

Year	Net Output	Employment	Investment	Capital Stock
1972	4679.44	350.900	312.70	5293.20
1973	5187.72	352.900	337.70	5543.00
1974	4933.58	351.000	387.30	5833.70
1975	4546.32	325.400	300.90	6026.30
1976	4550.35	312.300	271.00	6182.70
1977	4514.05	312.500	285.90	6337.20
1978	4606.83	314.200	316.10	6511.50
1979	4510.01	299.200	350.80	6711.40
1980	4034.00	237.500	342.00	6896.50
1981	3594.30	265.000	245.00	6957.90
1982	3669.37	255.000	253.00	7017.30
1983	3787.90	240.700	238.10	7047.10
1984	3836.30	234.200	283.00	7114.00

SIC 25 Chemical Industry

Year	Net Output	Employment	Investment	Capital Stock
1948	1278.60	321.800	195.90	3719.90
1949	1404.57	327.100	237.20	3908.00
1950	1568.33	340.000	301.80	4160.80
1951	1643.91	352.100	362.90	4475.40
1952	1517.94	353.000	334.70	4761.30
1953	1713.19	351.100	362.20	5071.00
1954	1883.25	361.300	399.10	5413.30
1955	1977.73	374.500	373.80	5725.00
1956	2110.00	379.400	519.50	6177.10
1957	2229.67	383.600	585.40	6694.50
1958	2292.65	384.500	627.70	7254.20
1959	2538.30	389.600	552.70	7738.20
1960	2802.83	401.000	490.20	8158.50
1961	2840.62	404.200	609.40	8693.70
1962	2954.00	399.400	603.00	9219.10
1963	3174.44	397.900	488.20	9627.50
1964	3495.67	396.700	618.90	10162.50
1965	3735.01	404.900	807.20	10882.10
1966	3949.16	413.400	931.30	11719.10
1967	4175.91	406.700	752.50	12356.20
1968	4528.62	386.700	765.30	12983.60
1969	4780.56	402.100	875.00	13706.50
1970	5038.80	402.100	1131.00	14685.70
1971	5139.58	397.600	990.00	15528.70
1972	5429.31	388.700	695.80	16075.60
1973	6071.75	390.300	520.80	16441.90
1974	6399.28	399.300	587.00	16866.10

(continued)

Appendix *(continued)*

Year	Net Output	Employment	Investment	Capital Stock
1975	5731.64	394.800	761.50	17449.50
1976	6474.86	387.000	855.80	18103.40
1977	6695.31	396.800	844.00	18721.50
1978	6770.89	401.200	1013.50	19490.50
1979	6959.84	392.200	962.60	20191.40
1980	6298.50	387.100	803.00	20691.30
1981	6317.40	359.600	573.50	20918.00
1982	6348.89	346.400	549.20	21106.60
1983	6815.00	325.300	588.30	21303.90
1984	7237.00	325.400	657.40	21532.90

SIC 26 Man-made Fibres

Year	Net Output	Employment	Investment	Capital Stock
1948	39.10	33.700	51.20	381.00
1949	47.00	36.200	64.10	443.50
1950	58.30	38.900	65.60	507.00
1951	60.20	37.500	65.00	569.40
1952	43.20	30.200	48.10	614.40
1953	63.20	32.600	46.30	656.30
1954	66.00	33.300	56.00	707.50
1955	69.70	31.500	60.20	762.30
1956	69.30	30.300	51.30	807.70
1957	70.90	30.300	70.00	870.90
1958	59.40	27.100	44.40	907.70
1959	73.30	28.500	44.30	943.40
1960	91.80	31.100	51.90	986.00
1961	92.00	31.200	57.60	1033.10
1962	106.40	31.600	58.70	1080.60
1963	126.00	31.700	52.50	1121.10
1964	145.50	34.100	63.80	1172.30
1965	158.80	35.600	72.40	1231.50
1966	173.40	35.800	79.60	1297.40
1967	187.00	35.100	77.60	1361.90
1968	242.50	33.700	86.90	1436.30
1969	256.70	36.600	75.60	1499.80
1970	279.50	39.800	134.60	1622.60
1971	296.60	38.400	98.70	1709.90
1972	299.40	35.700	62.70	1761.10
1973	355.30	35.700	112.40	1861.50
1974	305.90	38.700	77.30	1925.10
1975	275.70	34.000	56.90	1965.70
1976	303.10	34.100	40.30	1982.40
1977	270.30	31.700	30.10	1985.40
1978	289.90	29.700	30.90	1987.00

Appendix *(continued)*

Year	Net Output	Employment	Investment	Capital Stock
1979	277.60	34.700	49.80	2005.30
1980	202.60	26.600	39.00	2005.80
1981	172.00	17.500	17.50	1982.10
1982	138.00	16.300	13.70	1949.60
1983	158.20	14.800	13.60	1910.40
1984	159.60	15.100	12.30	1863.00

SIC 31 Metal Goods Nes

Year	Net Output	Employment	Investment	Capital Stock
1948	3757.63	503.800	117.50	2151.00
1949	3761.63	494.600	123.60	2240.60
1950	3833.74	495.500	133.50	2339.00
1951	4086.12	507.700	126.70	2430.00
1952	3933.89	503.700	117.80	2512.10
1953	3673.50	486.600	121.90	2597.10
1954	3893.83	492.900	141.40	2700.10
1955	4334.49	512.100	126.30	2785.10
1956	4238.35	516.100	153.10	2893.70
1957	4254.37	512.100	152.70	2999.80
1958	4146.21	505.600	152.00	3103.30
1959	4142.20	506.700	189.20	3242.90
1960	4691.03	547.100	202.70	3392.90
1961	4482.71	563.700	219.90	3555.00
1962	4330.49	558.800	183.90	3678.00
1963	4462.68	552.500	164.40	3779.30
1964	5163.73	575.900	179.00	3893.80
1965	5356.02	596.400	205.80	4035.30
1966	5063.58	606.100	221.40	4190.60
1967	4851.27	580.800	213.10	4328.80
1968	5287.92	582.700	227.50	4469.00
1969	5568.34	590.500	245.50	4617.70
1970	5488.22	593.400	309.90	4826.10
1971	5183.76	576.100	274.40	4999.40
1972	5275.90	549.400	219.90	5119.80
1973	5584.36	562.100	248.60	5264.60
1974	5388.07	565.300	281.90	5438.20
1975	4379.31	531.600	264.10	5587.10
1976	4923.37	504.700	242.00	5703.60
1977	5071.60	516.300	238.20	5804.60
1978	4963.43	519.700	264.60	5924.90
1979	4875.30	516.200	304.10	6078.80
1980	4006.00	489.900	232.00	6153.60
1981	3709.56	414.400	161.90	6143.00

(continued)

Appendix *(continued)*

Year	Net Output	Employment	Investment	Capital Stock
1982	3713.66	395.400	142.20	6110.90
1983	3785.82	364.700	138.50	6066.70
1984	4010.60	379.100	194.20	6067.60

SIC 32 Mechanical Engineering

1948	4695.44	897.500	125.70	3217.80
1949	5220.39	904.500	133.10	3304.90
1950	5793.95	919.400	151.90	3409.50
1951	6182.81	940.900	207.20	3568.40
1952	6134.20	963.800	213.90	3733.90
1953	6007.82	956.100	201.20	3885.10
1954	6377.24	968.000	201.50	4034.40
1955	6746.65	1005.800	233.90	4211.30
1956	6795.26	1034.700	245.10	4394.10
1957	7038.29	1038.100	252.70	4581.20
1958	6746.65	1034.700	253.70	4765.80
1959	6892.47	1006.800	227.10	4922.10
1960	7495.20	1055.700	287.70	5132.10
1961	8127.09	1105.800	305.80	5350.80
1962	8078.48	1118.500	280.00	5537.50
1963	8059.04	1081.400	276.20	5717.50
1964	8827.03	1098.400	328.10	5949.80
1965	9332.54	1147.300	336.60	6191.70
1966	9906.11	1169.400	354.00	6449.20
1967	9954.71	1153.500	350.10	6687.10
1968	10440.80	1130.200	380.50	6935.20
1969	11004.60	1160.900	356.20	7143.00
1970	11257.40	1183.200	423.30	7408.70
1971	10683.80	1125.200	356.70	7608.90
1972	10149.10	1047.900	331.30	7787.20
1973	11199.10	1039.700	420.70	8048.60
1974	11918.40	1052.800	467.50	8350.80
1975	11704.60	1042.300	444.10	8620.70
1976	11237.90	1012.200	429.50	8864.30
1977	11169.90	1011.900	470.80	9129.40
1978	10888.00	1024.700	531.70	9441.90
1979	10528.30	1010.500	505.60	9716.70
1980	9721.40	986.400	450.00	9919.20
1981	8671.49	888.900	341.30	9985.70
1982	8788.14	838.200	320.70	10030.50
1983	8496.50	767.200	314.50	10058.90
1984	3467.34	772.400	330.60	10094.40

Appendix *(continued)*

Year	Net Output	Employment	Investment	Capital Stock
SIC 33 Office Machinery, Data Processing Equipment				
1954	84.77	44.400	5.60	96.30
1955	83.73	46.900	7.80	102.80
1956	87.14	49.600	8.20	109.60
1957	75.85	52.400	8.70	116.80
1958	91.43	55.400	9.30	124.60
1959	90.00	57.000	9.30	132.30
1960	126.77	62.000	10.00	140.50
1961	137.75	67.000	10.00	148.50
1962	145.34	69.000	11.40	157.80
1963	159.62	68.000	13.10	168.60
1964	177.13	68.000	17.90	184.40
1965	188.75	71.000	22.10	204.30
1966	199.54	73.000	26.40	228.50
1967	207.05	70.000	25.90	251.70
1968	309.79	69.000	29.90	278.60
1969	352.73	76.000	37.10	312.10
1970	406.45	85.000	42.30	350.60
1971	306.00	86.000	30.50	377.30
1972	322.19	80.000	26.50	400.00
1973	419.00	75.000	16.80	412.70
1974	447.49	74.500	25.60	434.00
1975	484.36	68.000	28.60	457.90
1976	490.23	67.000	23.70	476.40
1977	555.60	66.000	32.20	502.50
1978	651.13	68.000	37.30	532.90
1979	779.34	72.000	40.10	565.60
1980	838.00	64.000	40.00	597.20
1981	716.49	73.000	28.70	615.40
1982	607.00	76.000	37.00	641.00
1983	1212.60	74.000	56.90	685.50
1984	1736.30	72.700	51.90	724.30
SIC 34 Electrical and Electronic Engineering				
1948	1418.05	446.600	75.30	1488.50
1949	1497.14	457.200	78.30	1545.80
1950	1627.09	475.100	77.30	1601.50
1951	1638.38	505.300	92.30	1671.80
1952	1740.07	539.900	105.60	1755.40
1953	1830.47	631.100	101.60	1834.10
1954	2022.55	563.300	110.40	1920.70
1955	2237.24	623.700	147.20	2041.80

(continued)

Appendix *(continued)*

Year	Net Output	Employment	Investment	Capital Stock
1956	2214.65	615.500	156.00	2169.30
1957	2367.18	628.200	143.50	2282.80
1958	2508.42	644.900	143.30	2394.60
1959	2762.66	659.100	145.20	2507.40
1960	2903.90	706.300	168.80	2640.70
1961	3016.89	728.500	162.40	2763.30
1962	3152.47	746.100	151.80	2872.30
1963	3321.97	753.000	175.70	3003.80
1964	3598.79	792.100	207.80	3167.20
1965	3531.00	800.100	228.80	3352.20
1966	3847.38	821.400	231.30	3538.50
1967	4107.26	809.100	223.00	3709.30
1968	4293.70	804.900	260.20	3907.50
1969	4632.67	815.100	265.60	4102.70
1970	4706.11	809.900	247.70	4275.20
1971	4785.21	795.300	247.20	4447.00
1972	5022.50	773.300	257.30	4630.10
1973	5711.75	793.500	304.50	4855.70
1974	5881.23	828.700	349.60	5122.50
1975	5581.81	766.700	305.90	5340.00
1976	5468.81	727.800	264.10	5508.60
1977	5621.35	740.300	304.10	5705.70
1978	5807.79	745.700	358.60	5949.50
1979	5756.94	748.600	376.30	6204.40
1980	5649.60	735.900	359.00	6430.90
1981	5259.78	663.500	329.90	6609.00
1982	5406.67	632.100	336.20	6789.40
1983	5915.10	619.400	387.80	7011.60
1984	6395.34	648.600	454.20	7292.60

SIC 341 Electronics

Year	Net Output	Employment	Investment	Capital Stock
1954	422.44	143.000	20.00	339.00
1955	481.54	169.000	24.00	359.00
1956	474.97	154.000	27.00	381.00
1957	503.42	164.000	24.00	400.00
1958	566.90	165.000	24.00	419.00
1959	641.32	173.000	26.00	438.00
1960	641.32	189.000	30.00	461.00
1961	689.47	192.000	28.00	482.00
1962	816.42	210.000	27.00	501.00
1963	853.63	213.000	28.00	521.00
1964	960.88	234.000	39.00	552.00
1965	1087.83	232.000	43.00	587.00

Appendix *(continued)*

Year	Net Output	Employment	Investment	Capital Stock
1966	1146.93	241.000	43.00	623.00
1967	1210.41	245.000	43.00	658.00
1968	1409.59	259.000	47.00	693.00
1969	1346.11	277.000	52.00	733.00
1970	1540.92	272.000	55.00	775.00
1971	1523.40	270.000	51.00	813.00
1972	1665.68	268.000	58.00	858.00
1973	1891.12	285.000	68.00	915.00
1974	1950.22	303.000	84.00	986.00
1975	1726.96	272.000	62.00	1034.00
1976	1775.12	264.000	59.00	1077.00
1977	1948.03	272.000	67.00	1129.00
1978	2079.36	272.000	85.00	1199.00
1979	2195.37	268.000	95.20	1278.00
1980	2188.80	265.000	93.70	1354.00
1981	2182.23	245.500	83.30	1418.00
1982	2329.13	239.000	66.80	1465.00
1983	2563.00	244.900	66.80	1465.00
1984	2961.40	257.400	66.80	1465.00

SIC 342 Electrical Engineering

Year	Net Output	Employment	Investment	Capital Stock
1954	1600.11	420.300	90.40	1581.70
1955	1755.70	454.700	123.20	1682.80
1956	1739.68	461.500	129.00	1788.30
1957	1863.76	464.200	119.50	1882.80
1958	1941.52	479.900	119.30	1975.60
1959	2121.34	486.100	119.20	2069.40
1960	2262.58	517.300	138.80	2179.70
1961	2327.42	536.500	134.40	2281.30
1962	2336.05	536.100	124.80	2371.30
1963	2468.34	540.000	147.70	2482.80
1964	2637.91	558.100	168.80	2615.20
1965	2443.17	568.100	185.80	2765.20
1966	2700.45	580.400	188.30	2915.50
1967	2896.85	564.100	180.00	3051.30
1968	2884.11	545.900	213.20	3214.50
1969	3286.56	538.100	213.60	3369.70
1970	3165.19	537.900	192.70	3500.20
1971	3261.81	525.300	196.20	3634.00
1972	3356.82	505.300	199.30	3772.10
1973	3820.63	508.500	236.50	3940.70
1974	3931.01	525.700	265.60	4136.50

(continued)

Appendix *(continued)*

Year	Net Output	Employment	Investment	Capital Stock
1975	3854.85	494.700	243.90	4306.00
1976	3693.69	463.800	205.10	4431.60
1977	3673.32	468.300	237.10	4576.70
1978	3728.43	473.700	273.60	4750.50
1979	3561.57	480.600	281.10	4926.40
1980	3460.80	470.900	265.30	5076.90
1981	3077.55	418.000	246.60	5191.00
1982	3086.54	393.100	269.40	5324.40
1983	3352.10	374.500	321.00	5546.60
1984	3433.94	391.200	387.40	5827.60

SIC 35 Motor Vehicles and Parts

Year	Net Output	Employment	Investment	Capital Stock
1948	1234.28	284.200	104.30	2438.90
1949	1454.69	295.700	93.30	2470.10
1950	1743.22	313.100	135.10	2569.50
1951	1747.23	321.500	151.20	2704.60
1952	1775.28	333.200	145.00	2827.20
1953	1983.66	330.700	128.70	2924.50
1954	2260.17	351.900	136.40	3019.80
1955	2620.84	377.500	185.30	3157.80
1956	2280.21	362.000	225.80	3330.50
1957	2496.61	350.400	229.10	3505.20
1958	2789.15	362.200	173.60	3625.60
1959	3274.05	378.800	227.60	3796.10
1960	3879.16	429.900	256.70	3982.40
1961	3494.45	408.900	380.30	4280.60
1962	3710.85	421.900	351.90	4547.90
1963	4323.98	442.000	356.30	4816.60
1964	4828.92	471.000	354.10	5076.60
1965	4856.97	485.100	384.90	5347.50
1966	4792.85	487.100	402.20	5623.00
1967	4540.38	458.000	438.40	5896.50
1968	5061.35	462.000	364.90	6086.30
1969	5353.89	488.100	406.00	6298.80
1970	5361.90	507.100	511.70	6601.20
1971	5229.66	503.100	338.40	6720.10
1972	5337.86	488.900	182.80	6686.10
1973	5586.32	509.300	307.70	6779.50
1974	5153.52	496.000	393.00	6953.20
1975	4672.63	456.400	300.00	7033.00
1976	4716.71	447.000	257.10	7084.40
1977	5089.40	463.200	341.00	7231.00
1978	4764.80	469.800	417.80	7444.20

Appendix *(continued)*

Year	Net Output	Employment	Investment	Capital Stock
1979	4636.56	433.100	648.00	7873.40
1980	4007.40	411.800	539.00	8189.10
1981	3322.10	354.700	362.90	8327.20
1982	3209.90	319.000	254.10	8365.80
1983	3362.20	306.900	341.10	8459.20
1984	3258.00	289.700	400.10	8599.60

SIC 36 Other Transport Equipment

Year	Net Output	Employment	Investment	Capital Stock
1948	4360.24	621.100	40.60	3400.10
1949	4296.84	622.400	54.60	3438.00
1950	4511.68	600.400	79.70	3500.40
1951	4511.68	603.400	170.90	3652.80
1952	4008.04	643.600	217.10	3850.70
1953	4029.17	663.900	199.00	4026.20
1954	4712.44	685.900	121.40	4121.80
1955	4839.23	701.800	92.50	4184.70
1956	4733.57	720.400	112.80	4265.10
1957	4708.91	718.500	116.40	4346.20
1958	4642.00	694.200	95.90	4406.30
1959	4550.42	665.200	93.10	4462.70
1960	4391.93	652.700	96.60	4515.50
1961	4057.34	639.900	106.20	4572.00
1962	3912.94	606.100	97.70	4614.80
1963	3638.23	554.200	63.60	4619.40
1964	3419.86	527.400	68.60	4623.60
1965	3638.23	507.200	66.00	4622.10
1966	3617.09	496.100	95.10	4646.00
1967	3557.22	491.700	107.50	4641.80
1968	3557.22	467.900	123.00	4635.60
1969	3536.09	459.400	132.20	4616.30
1970	3247.28	444.700	124.20	4573.10
1971	3458.60	432.700	73.10	4475.20
1972	3384.64	412.300	81.00	4387.40
1973	3539.61	406.900	97.90	4320.10
1974	3631.18	410.800	131.90	4289.60
1975	3469.17	409.500	152.10	4279.60
1976	3310.68	403.900	119.00	4231.40
1977	3169.80	390.900	93.00	4147.80
1978	3374.08	388.200	117.00	4078.20
1979	3254.33	432.200	145.10	4028.60
1980	3522.00	406.900	154.00	3975.20
1981	3645.27	365.100	125.00	3888.00

(continued)

Appendix *(continued)*

Year	Net Output	Employment	Investment	Capital Stock
1982	3539.61	344.300	130.00	3839.20
1983	3345.86	324.900	135.00	3802.60
1984	3208.50	292.600	151.00	3791.90

SIC 37 Instrument Engineering

Year	Net Output	Employment	Investment	Capital Stock
1948	181.90	96.700	8.90	220.30
1949	191.40	102.900	13.40	230.50
1950	207.60	107.800	13.00	240.30
1951	209.50	107.600	10.90	247.90
1952	221.90	103.300	10.20	254.90
1953	234.30	106.600	9.80	261.30
1954	258.10	108.200	7.70	265.40
1955	285.70	110.400	10.70	272.30
1956	282.90	112.000	10.50	278.50
1957	302.90	114.100	10.80	284.90
1958	320.90	113.700	10.60	290.70
1959	353.30	112.100	11.40	297.30
1960	371.40	121.100	15.00	307.00
1961	394.30	129.400	17.60	318.60
1962	420.00	129.400	16.90	329.00
1963	448.60	131.000	18.80	341.30
1964	499.10	127.700	23.90	358.60
1965	563.80	132.700	21.40	373.40
1966	584.80	132.700	31.70	398.60
1967	600.00	135.100	31.00	422.00
1968	671.40	131.000	35.70	448.80
1969	720.00	129.400	36.70	475.60
1970	780.00	132.700	42.30	507.40
1971	806.70	135.500	38.90	536.00
1972	796.20	128.200	39.80	565.90
1973	859.10	131.000	48.00	604.00
1974	918.10	131.100	55.40	649.30
1975	927.60	127.400	44.70	683.30
1976	908.60	122.300	37.10	708.60
1977	930.50	122.500	42.50	738.10
1978	986.70	123.000	45.40	769.70
1979	974.30	126.300	44.50	799.40
1980	952.40	122.800	43.00	826.90
1981	972.40	111.000	35.20	844.50
1982	902.90	108.900	37.80	864.40
1983	903.60	103.600	41.50	887.20
1984	964.80	108.800	45.70	913.20

Appendix *(continued)*

Year	Net Output	Employment	Investment	Capital Stock
SIC 411/22 Food				
1948	3557.92	284.900	158.80	1841.00
1949	3760.87	295.700	166.90	1954.40
1950	3982.84	313.100	202.90	2103.50
1951	3856.00	321.500	187.20	2237.50
1952	4039.92	333.200	177.80	2362.70
1953	4376.05	330.700	172.30	2483.00
1954	4363.37	351.900	193.90	2623.90
1955	4407.76	377.500	198.30	2768.40
1956	4534.60	362.000	214.20	2927.70
1957	4540.94	350.400	223.00	3094.70
1958	4705.84	362.200	218.30	3265.10
1959	4781.94	378.800	217.10	3415.60
1960	4883.42	429.900	241.90	3600.10
1961	5022.94	408.900	243.30	3784.00
1962	5181.50	421.900	256.50	3981.10
1963	5244.92	442.000	260.40	4182.70
1964	5232.23	471.000	281.50	4404.90
1965	5340.05	485.100	260.60	4604.30
1966	5479.57	487.100	268.20	4809.20
1967	5530.31	458.000	318.30	5066.10
1968	5745.94	462.000	363.60	5368.10
1969	5879.13	488.100	354.20	5657.80
1970	5910.84	507.100	367.90	5962.70
1971	5910.84	632.800	327.90	6227.80
1972	6145.50	620.900	336.30	6499.00
1973	6278.68	619.600	379.10	6805.10
1974	6164.52	625.700	430.00	7154.20
1975	5917.18	589.900	387.70	7450.80
1976	6101.10	581.700	362.00	7703.50
1977	6240.63	581.800	436.70	8014.70
1978	6291.36	575.300	463.90	8343.50
1979	6373.81	574.800	468.00	8667.00
1980	6342.10	566.100	433.00	8939.20
1981	6285.02	539.800	406.00	9167.40
1982	6500.65	521.800	427.00	9402.40
1983	6557.73	495.600	406.00	9587.40
1984	6633.84	485.000	492.90	9833.50
SIC 424/9 Drink and Tobacco				
1948	1299.71	141.200	59.70	672.00
1949	1305.94	145.000	65.20	717.80

(continued)

Appendix *(continued)*

Year	Net Output	Employment	Investment	Capital Stock
1950	1277.89	141.800	65.50	763.70
1951	1293.47	140.800	69.20	813.60
1952	1277.89	138.100	66.10	860.60
1953	1277.89	137.800	56.50	898.20
1954	1312.17	136.800	48.90	927.80
1955	1383.86	135.600	52.00	960.20
1956	1402.56	136.600	54.80	994.90
1957	1461.78	140.400	66.70	1041.20
1958	1480.48	141.300	77.10	1097.50
1959	1580.22	144.900	80.90	1157.40
1960	1645.67	145.900	105.40	1241.90
1961	1711.12	148.800	117.60	1337.90
1962	1701.77	146.800	115.10	1431.40
1963	1804.63	147.900	119.90	1529.90
1964	1935.53	146.500	129.90	1638.20
1965	1991.64	146.800	159.20	1775.20
1966	2069.56	146.800	134.10	1886.30
1967	2131.89	146.800	157.10	2021.20
1968	2222.28	139.500	169.80	2168.80
1969	2331.37	140.000	158.10	2303.70
1970	2403.05	139.700	217.60	2498.70
1971	2459.16	141.200	246.10	2722.10
1972	2549.54	139.800	202.80	2901.40
1973	2798.89	139.800	278.00	3153.70
1974	2854.99	145.600	283.10	3408.50
1975	2858.11	143.300	216.10	3592.50
1976	2939.14	139.700	196.00	3750.00
1977	2929.79	139.200	239.90	3944.80
1978	3110.57	138.300	247.90	4143.30
1979	3176.02	140.600	222.10	4312.40
1980	3116.80	140.900	239.00	4493.80
1981	3007.71	126.300	180.10	4608.50
1982	2932.90	122.200	170.00	4707.80
1983	2976.50	115.400	176.00	4800.20
1984	3020.20	125.500	164.00	4867.50

SIC 43 Textiles

Year	Net Output	Employment	Investment	Capital Stock
1948	2078.40	767.500	191.50	4191.50
1949	2251.10	795.900	237.20	4301.80
1950	2463.30	832.600	244.60	4419.60
1951	2467.50	852.500	240.70	4533.20
1952	2001.40	757.500	180.50	4581.50
1953	2371.80	798.700	172.40	4618.40

Appendix *(continued)*

Year	Net Output	Employment	Investment	Capital Stock
1954	2473.70	319.500	209.10	4687.00
1955	2413.40	792.200	215.30	4762.70
1956	2384.20	777.200	181.30	4805.30
1957	2388.40	769.200	171.90	4838.80
1958	2153.30	715.400	141.10	4841.10
1959	2271.90	701.600	141.90	4840.80
1960	2378.00	703.300	193.90	4891.30
1961	2294.80	695.200	258.00	4999.10
1962	2253.20	660.600	216.10	5064.00
1963	2365.50	645.800	215.10	5125.80
1964	2509.10	644.100	275.20	5247.20
1965	2581.90	631.100	327.50	5420.10
1966	2575.70	624.200	309.30	5575.30
1967	2521.60	578.200	221.70	5657.30
1968	2904.40	567.800	336.50	5865.30
1969	2993.80	577.300	373.00	6121.20
1970	2948.10	549.600	281.60	6294.80
1971	2923.10	503.000	241.00	6436.90
1972	2964.70	484.100	210.40	6552.60
1973	3104.10	480.800	301.10	6760.60
1974	2842.00	468.200	370.50	7037.00
1975	2615.20	421.400	225.10	7161.60
1976	2696.30	406.900	179.00	7218.90
1977	2717.10	409.000	156.00	7246.30
1978	2625.60	391.600	174.50	7290.10
1979	2517.40	375.900	167.60	7322.90
1980	2080.50	328.500	114.00	7281.70
1981	1909.90	276.200	83.30	7205.60
1982	1864.10	263.300	90.40	7124.70
1983	1899.50	251.500	97.20	7027.70
1984	1951.50	231.100	133.00	6939.50

SIC 4445 Leather, Footwear and Clothing

Year	Net Output	Employment	Investment	Capital Stock
1948	1940.70	654.000	22.40	949.10
1949	2090.20	682.300	48.80	962.20
1950	2198.70	697.600	47.80	972.20
1951	2111.90	701.800	39.20	972.50
1952	1993.70	702.500	33.10	964.40
1953	2162.50	646.800	38.20	961.40
1954	2181.80	673.000	34.00	956.40
1955	2210.70	668.500	35.90	953.80
1956	2193.80	654.900	34.60	949.80

(continued)

Appendix *(continued)*

Year	Net Output	Employment	Investment	Capital Stock
1957	2150.40	642.400	34.60	945.70
1958	2015.40	610.900	31.60	942.30
1959	2191.40	602.800	36.80	937.70
1960	2295.10	619.700	47.40	948.80
1961	2304.70	619.900	53.40	965.90
1962	2196.20	612.400	50.50	982.00
1963	2196.20	595.200	42.90	992.70
1964	2316.80	592.300	56.90	1019.40
1965	2401.20	584.400	60.30	1053.00
1966	2398.70	580.100	55.10	1084.10
1967	2302.30	546.000	50.20	1111.80
1968	2430.10	539.100	62.50	1152.40
1969	2381.90	543.400	63.30	1189.00
1970	2345.70	517.400	57.80	1220.90
1971	2495.20	513.400	55.40	1250.70
1972	2538.60	508.100	64.00	1289.60
1973	2680.80	500.400	82.40	1342.40
1974	2622.90	483.700	77.70	1388.00
1975	2627.80	459.900	65.80	1419.40
1976	2562.70	439.500	54.80	1437.50
1977	2700.10	445.400	56.10	1450.50
1978	2736.30	432.500	67.30	1471.70
1979	2777.20	432.900	71.50	1493.80
1980	2410.80	393.900	60.00	1506.90
1981	2251.70	341.600	35.70	1488.60
1982	2234.80	323.400	59.70	1492.00
1983	2348.10	315.200	61.00	1496.10
1984	2447.00	294.500	72.00	1510.60

SIC 46 Timber and Wooden Furniture

Year	Net Output	Employment	Investment	Capital Stock
1948	910.455	274.800	23.60	208.50
1949	1035.89	282.300	38.40	241.50
1950	1089.95	285.500	40.40	276.20
1951	1174.29	289.300	38.20	309.30
1952	1040.21	274.500	26.40	330.20
1953	1122.39	275.900	24.20	348.30
1954	1280.26	278.300	30.10	371.30
1955	1321.35	281.500	33.20	398.30
1956	1275.93	271.900	26.10	417.10
1957	1340.81	268.500	24.40	433.90
1958	1319.19	263.400	27.80	453.80
1959	1505.17	268.400	38.90	484.20
1960	1567.89	276.300	35.80	511.60

Appendix *(continued)*

Year	Net Output	Employment	Investment	Capital Stock
1961	1624.11	275.300	30.10	533.90
1962	1591.67	272.300	25.30	551.40
1963	1654.39	269.400	34.50	578.70
1964	2173.41	276.200	38.90	611.10
1965	2195.04	283.100	36.20	641.60
1966	2147.46	278.200	30.00	666.60
1967	2214.50	266.400	34.30	694.80
1968	2385.35	284.100	62.50	750.80
1969	2233.97	272.300	44.90	784.30
1970	2242.62	260.500	51.60	822.60
1971	2309.66	260.800	64.80	873.20
1972	2543.22	265.500	80.70	938.10
1973	2971.41	280.800	119.70	1037.40
1974	2528.08	273.100	113.00	1128.20
1975	2495.64	253.200	77.70	1181.90
1976	2549.71	253.400	74.10	1229.80
1977	2376.70	248.200	84.40	1282.70
1978	2436.99	246.100	83.10	1332.50
1979	2523.75	248.900	103.20	1401.60
1980	2162.60	234.800	106.00	1475.90
1981	1948.50	216.300	53.20	1493.60
1982	1916.06	201.200	54.70	1512.10
1983	2073.90	205.500	69.00	1543.50
1984	2082.60	202.200	57.50	1562.20

SIC 47 Paper, Printing and Publishing

Year	Net Output	Employment	Investment	Capital Stock
1948	2715.97	447.000	125.70	2812.30
1949	3107.57	465.600	127.20	2867.40
1950	3537.07	489.500	138.90	2933.50
1951	3676.03	497.700	144.80	3006.20
1952	3012.83	494.200	121.10	3055.10
1953	3379.17	484.900	100.90	3083.00
1954	3960.26	506.300	138.70	3148.80
1955	4269.75	525.200	154.90	3231.00
1956	4212.91	536.600	193.10	3352.70
1957	4307.65	545.400	217.90	3499.70
1958	4446.60	546.800	191.40	3620.60
1959	4648.72	550.100	181.60	3732.00
1960	5103.49	578.300	203.10	3863.30
1961	5015.06	594.400	231.20	4019.60
1962	5008.75	602.400	225.50	4169.60
1963	5160.35	601.400	236.00	4330.10

(continued)

Appendix *(continued)*

Year	Net Output	Employment	Investment	Capital Stock
1964	5634.05	602.400	274.70	4527.60
1965	5754.06	612.500	300.00	4749.10
1966	5911.96	619.500	275.00	4942.90
1967	5886.70	608.500	277.10	5143.00
1968	6139.35	609.500	283.00	5351.60
1969	6366.73	616.500	282.20	5560.80
1970	6360.41	622.500	308.20	5799.30
1971	6088.82	592.600	268.60	6000.50
1972	6373.05	576.100	245.80	6180.20
1973	6922.56	571.200	295.60	6408.10
1974	6909.92	585.700	333.40	6671.20
1975	5975.13	562.400	295.20	6893.90
1976	6158.29	538.900	258.90	7071.70
1977	6429.89	534.200	295.00	7279.90
1978	6556.22	537.600	351.20	7540.80
1979	6821.50	547.100	413.10	7862.30
1980	6316.20	541.300	414.00	8176.90
1981	6000.40	511.800	322.10	8395.60
1982	5791.90	492.600	291.00	8575.70
1983	5817.20	473.300	325.00	8773.40
1984	6006.70	482.100	364.00	8994.20

SIC 4849 Rubber, Plastics and Other Manufacturing

Year	Net Output	Employment	Investment	Capital Stock
1948	966.67	239.000	68.70	781.80
1949	960.03	235.900	64.00	822.90
1950	1119.48	246.700	59.70	858.00
1951	1189.24	267.000	59.80	892.80
1952	1056.36	246.500	54.80	922.40
1953	1152.70	249.600	43.40	940.10
1954	1325.44	268.600	62.60	976.20
1955	1454.99	281.800	100.30	1049.40
1956	1375.27	280.800	109.30	1130.60
1957	1448.35	282.500	102.70	1204.20
1958	1415.13	281.200	106.20	1281.90
1959	1574.58	282.700	113.50	1366.20
1960	1797.15	305.100	128.30	1466.40
1961	1783.86	311.500	136.90	1577.20
1962	1873.55	310.400	139.80	1692.50
1963	1996.46	313.600	141.60	1812.30
1964	2308.72	327.500	198.00	1990.60
1965	2451.56	357.100	406.00	2377.90
1966	2554.54	343.500	427.40	2786.40
1967	2660.84	330.700	403.60	3169.20

Appendix *(continued)*

Year	Net Output	Employment	Investment	Capital Stock
1968	2986.39	346.700	506.80	3654.30
1969	3102.65	357.400	627.80	4253.50
1970	3119.26	361.600	264.60	4488.70
1971	3115.94	353.100	274.70	4730.60
1972	3255.46	352.900	250.70	4945.30
1973	3640.80	366.700	279.20	5177.70
1974	3554.43	375.300	274.10	5397.50
1975	3238.85	349.400	215.60	5551.80
1976	3507.93	348.500	235.00	5719.50
1977	3733.82	354.000	250.60	5886.20
1978	3796.93	351.900	259.00	6053.50
1979	3803.58	345.900	254.70	6209.30
1980	3321.90	322.100	249.00	6355.80
1981	3039.54	235.800	180.80	6389.10
1982	2896.70	269.300	194.30	6404.80
1983	2851.52	256.800	196.00	6392.10
1984	3039.87	246.200	219.50	6371.90

[a] SIC 33, 341 and 342: 1948–53 not available.
Source: CSO and own estimates.

Net Output

Index of Output of the Production Industries (1980) supplied by the Central Statistical Office for the following years:

SIC 21 and 22 1967 to 1984
SIC 23 and 24 1948 to 1984
SIC 25 1948 to 1984
SIC 26 1948 to 1984
SIC 31 1967 to 1984
SIC 32 1948 to 1984
SIC 33 1973 to 1984
SIC 34 1973 to 1984
SIC 35 1948 to 1984
SIC 36 1948 to 1984
SIC 37 1973 to 1984
SIC 411 to 422 1948 to 1984
SIC 424 to 429 1948 to 1984
SIC 43 1948 to 1984
SIC 44 and 45 1948 to 1984
SIC 46 1973 to 1984

SIC 47 1948 to 1984
SIC 48 and 49 1973 to 1984

The rest of the years are own estimates based on gross output figures supplied by the Institute of Employment Research at Warwick University – SICs 21, 22, 31, 34, 37, 46, 48, and 49. SIC 33 and the disaggregation of SIC 34 into Electronics and Electrical Engineering from Soete and Dosi (1983).

The 1980 figures to convert the index into pounds from the Census of Production.

Employment

Data on Employees in Employment in Great Britain supplied by the Statistics Division of the Department of Employment for 1971 to 1984 (June figures) for all the sectors except the disaggregation of SIC 34 (for that see Soete and Dosi, 1983).

Prior to 1971 estimates based on figures supplied by the Institute of Employment Research at Warwick University.

Capital Stock and Investment

Gross Capital Stock and Gross Capital Formation at 1980 prices supplied by the Central Statistical Office for all years for all the sectors except the disaggregation of SIC 34 (for that see Soete and Dosi, 1983).

9

The UK Tertiary
Service Sector

Ken Guy

The plan for this chapter is as follows: the first section introduces
the tertiary (service) sector in terms of definitions, scope and major
characteristics. We are adopting a very broad definition of the 'service
sector' as the entire tertiary sector, including non-manufacturing industries
such as construction, utilities, transport and communications. The second
section then reviews the empirical evidence on output and employment
trends in the nine major service subsectors. This is followed by a review
of various theories of service sector growth in the third section.

In the fourth section the evidence now available on the changing nature
of the capital stock of the service sectors is examined in relation to capital-
embodied technical change. The final section sums up the trends and their
likely implications for future employment.

Definition and Scope of the Tertiary Sector

Before it is possible to discuss either service sector growth or the role of
technology in this, it is necessary to describe and define the nature and
composition of the sector. The first point which needs to be emphasized
immediately is that however the tertiary sector is defined, its most
outstanding feature is its heterogeneous, 'ragbag' nature. Its constituent
components vary tremendously across a number of dimensions. This
heterogeneity hinders both the exposition of simple theories of service
sector growth and an analytically useful disaggregation of the component
parts of the sector.

There have been numerous attempts to provide analytical schemata of
this nature. Many of these have relied on descriptions which can be
variously classified as:

169

functional-relational;
organizational;
temporal;
empirical;
technological.

Functional-relational classifications sort out sectors on the basis of the nature of the activities carried out within them and the nature of the relationships existing between service producers and consumers. In the traditional three-sector economy composed of (primary) agricultural and extractive activities, (secondary) manufacturing activities, and tertiary or service activities, the service sector consists of everything other than agriculture, extraction and manufacturing. A more refined classification is the Browning–Singelmann (1978) system. This reclassifies construction and utilities into a transformative sector and then disaggregates the remaining service sectors into: a distributive sector concerned primarily with the distribution of goods; producer services which are largely intermediate services provided to sectors of the economy other than the final consumer; and social services, i.e. services largely provided collectively. Personal services constitute the residual in this scheme.

Organizational classifications typically make a distinction between publicly and privately provided services and have been popular largely because statistics were often available in this form. They are useful because growth mechanisms may be different in each sector as a consequence of organizational form.

Temporal classifications such as the one developed by Katouzian (1970) make a distinction between 'old' services such as public transport, cinemas, laundries, shoe repairs and domestic service, all sectors for which demand has generally decreased over time, and 'new' services like health and education which have seen an expansion of demand. In their dependence on economic indicators as markers, temporal classifications share much with empirical classifications as advocated by Smith (1972). These propose the use of measurable economic parameters such as a high ratio of value added to total inputs to distinguish service industries. Whilst laudable in their attempts to introduce some kind of precision into taxonomic endeavours, empirical classifications of this nature suffer from the fact that membership of a service sector cluster becomes a function of time and the growth rates of the economic indicators chosen to determine membership. Tracing the economic progress of any one cluster of sectors becomes problematical when the membership of the cluster is prone to change over time. Exactly the same criticism can be applied to technological classifications which are based on measures such as capital intensity. Low capital intensities have indeed been characteristic of many service sectors in the past, but this is now changing. Does a sector like banking and insurance stop being a service sector because its capital intensity growth brings it more in line with capital equipment levels in manufacturing?

The classification scheme used in this chapter resembles most closely the Browning–Singelmann scheme but draws upon most of the schemata outlined above. The levels of aggregation used were primarily determined by the availability of plant and machinery capital stock data in the 1980 SIC classification.

Nine separate service subsectors will be considered throughout this chapter. These nine subsectors have been regrouped into four distinct clusters on the basis of some of the criteria used in other classification schemes.

The construction sector is considered apart from the others largely because of its transformative function and the strong influence of public policy. Education and health and public administration (both central and local) are likewise grouped together under public social services to reflect both organizational and functional–relational criteria. Similarly, financial, distribution and 'other private' services are termed private commercial services. The financial and business services sector as defined here corresponds very closely to the producer services category used by Browning and Singelmann. Finally, the utilities, transport and communications sectors have been grouped together and termed network services. The rationale for this particular cluster is partly functional – all make use of infrastructural networks such as roads, railways, electricity grids and communication networks to provide services – and partly empirical and technological in that all three have markedly different plant and machinery capital stocks from other service sectors. In fact, when plant and machinery capital intensity ratios are examined later in this chapter, it will be seen that these figures alone can be used to distinguish and differentiate between the major service sector groupings outlined here.

Output and Employment in the Service Sector

The following section uses the above categories to trace the growth of service sector output and employment over the last 30 years or so. This is followed by a discussion of the various growth mechanisms which have been propounded, before the focus is shifted to changes in the capital stock of the service sectors. For the sake of convenience, although much of the data on time trends is presented for the three decades, 1953–63, 1963–73, 1973–83, these periods are referred to in the text as the fifties, sixties and seventies.

Employment Growth

Charting employment changes in the UK service sector over the last 30 years or so is no easy task. Employment time series are available from a number of sources, but changes in the Standard Industrial Classification (SIC) in 1948, 1958, 1968 and 1980 have made the collection of definitionally consistent series difficult. As yet, there is no official source of UK employment statistics based on SIC (1980) which extends back as far as the period of interest of this chapter. The series used in this chapter is therefore a mixture of an official time series of employees in employment based on SIC (1980) categories extending back as far as 1973, together with a self-estimated time series for the preceding years constructed by reallocating pre-SIC (1980) disaggregated data into more aggregated SIC (1980) categories. The ease with which this earlier time series was constructed varied from sector to sector, and inevitably a perfect reclassification was not possible. For example, the early figures for employment in the distribution sector do not include those employed in the repair industries, whereas later years do so. Where problems of reclassification for these earlier years are felt to influence some of the results given in this chapter, they are discussed in the text.

Perhaps more importantly, the employment time series constructed for this chapter take no account of self-employed people, nor do they adjust to the full-time equivalent of part-time employees. Again this is primarily because of the lack of adequate official time series for the years of interest. Self-employment in the service sector accounted for about 10 per cent of total service sector employment in 1983 (*Employment Gazette*, July 1984), and in the construction sector this amounted to around 30 per cent. When looking at time trends and growth rates these absolute differences often make little difference to the results obtained. Thus, for example, the numbers of self-employed showed variations in individual sectors between 1951 and 1971, rising in 'entertainment' and falling in 'hairdressing', but for the whole of the 'other private' services category the proportions remained the same at 17 per cent (Smith, 1986a). But for the last few years, there may have been a significant rise in the proportion of self-employed.

As defined in this chapter, the service industries accounted for some 15 million employees in employment in 1983, about 70 per cent of the total work force defined in this way. In 1953 the corresponding figures were around 10.5 million and 50 per cent. The service sector grew in absolute

Table 9.1 Employment growth in the service sectors

Sector	Employment growth rates (% p.a.)			
	1953–63	1963–73	1973–83	1953–83
Network services	−0.1	−0.8	−1.1	−0.8
Utilities	0.6	−1.1	−0.5	−0.3
Transport	−0.8	−1.8	−1.7	−1.4
Communications	1.3	0.9	−0.3	0.6
Private commercial services	2.7	0.5	1.3	1.5
Financial and business	2.8	4.0	2.4	3.1
Distribution	3.1	−0.1	0.6	1.2
Other private	1.5	−1.4	2.2	0.8
Construction	1.7	−1.8	−2.6	−0.9
Public social services	2.2	3.1	1.1	2.1
Education and health	4.1	3.9	1.9	3.3
Public administration	0.2	1.8	−0.4	0.5
All services	2.0	0.8	0.6	1.1
Manufacturing	0.6	−0.5	−3.5	−1.1

and relative terms over this period, although in absolute terms there was a decline from a peak in 1980 of over 15.3 million workers.

It should be noted, however, that this overall employment growth averaged about 1 per cent per annum. The fifties was the period of highest growth, but this slackened off in the sixties and seventies. Nevertheless, employment fared better than in the manufacturing sector in all periods and this led to a gradual shift to an economy dominated by service sector employment. Table 9.1 summarizes employment growth in the service sector from 1953 to 1983.

Of the nine service sectors defined in this chapter, two dominate employment. In 1983 the distribution sector accounted for over a quarter of all service employment, i.e. over 4 million. The number involved in education and health amounted to over 20 per cent – more than 3 million. Taken together, these two very different sectors accounted for half the work-force employed in the service sectors. Public administration and financial and business services each employed just over 10 per cent of the service sector work-force. Thus these four sectors accounted for about half of all UK employment.

Over the period 1953–83 there was no uniform pattern of growth across the whole service sector. As noted in the previous section, the service sector as defined here is extremely heterogeneous. Table 9.1 indicates that although the whole sector grew slowly over the 30-year period, some parts suffered employment decline akin to the manufacturing sector, whilst others grew relatively quickly and consistently. In terms of the four distinct groupings, the highly capital intensive network sectors (utilities, transport and communications) lost employment, though communications did in fact show a slight employment growth when considered in isolation. The construction sector was another to exhibit employment decline, but the other two groupings, the private commercial service grouping (distribution, financial and 'other private') and the public social service grouping (education and health and public administration), both experienced considerable expansion.

Just as employment change over the whole period differed from one sector to another, growth within and between sectors rarely proceeded at consistent rates over the decades. The fifties was generally a time of service sector expansion, but even during this period employment declined in the transport sector. The number of jobs in public administration remained static over the fifties, though a change in SIC categories probably disguised a slight rise, but the public social service grouping as a whole expanded fairly quickly because of the dramatic increase in employment in the education and health sector with the development of the Welfare State. In addition, the private commercial service grouping burgeoned, with financial and business services and distribution both growing very rapidly at around 3 per cent per annum. Many of the component industries of the 'other private' services sector grew appreciably too, but overall growth was retarded by the decline in domestic service. In direct contrast, the network sectors taken together lost employment whereas employment opportunities within the construction industries continued to expand with increased demand for housing and civil engineering projects.

The sixties saw manufacturing employment peak around the middle of the decade and then decline. During this period service sector employment continued to grow, but at a slower pace than in the previous decade. Again, however, this pace was not uniform. The network and construction groupings actually shed labour at a greater rate than manufacturing, whilst both the public social service and the private commercial service groupings continued to grow, the latter slowly, and the former more quickly. Even these aggregate trends disguised internal variations, for whilst transport and utilities suffered employment losses, employment in communications rose slightly. Similarly, although overall employment growth in the private commercial service grouping was low over the sixties, this is largely a reflection of the slightly negative employment performance of the large distribution sector associated with the expansion of 'self-service' shopping and the continued decline of domestic service in the 'other private' services sector. In the financial and business services sector, employment growth was healthy.

Although employment declined in some service sectors in the seventies, all sectors fared better than manufacturing did. The public service and private commercial service groupings changed positions with regard to employment growth in the sixties and seventies. Whereas the former grew faster than the latter in the sixties, the private commercial service sector overtook the public social sector in the following decade primarily because of changes in public policy. Education grew in the early seventies but in 1984 employment was no higher than in 1976. Employment in health services grew more slowly than previously, whilst public administration peaked in the mid-seventies and declined afterwards. The decline in construction was related both to cyclical factors and policy changes. By contrast, employment growth quickened considerably in the 'other private services' sector, continued at a fairly high rate in the financial and business service sector and reversed a decline in the distribution sector.

Output Measurement and Output Growth

Employment growth in the services is intimately linked to output expansion, but before sectoral trends in output can be delineated, some discussion of the measurement of service sector output is necessary.

There are no simple, uniform measures of output applicable to all service sectors. Smith (1972) notes that four measures are commonly used: physical quantity measures, such as passenger journeys in the transport sector; deflated money values of output, such as the volume of sales in the distribution sector; deflated wage bills, as in the health sector; and, finally, the number of employees, a measure frequently used in services such as public administration and education.

The CSO index numbers of output at constant factor cost used in this chapter are derived using a combination of the measures given above. For each sector the combination is different. Thus, the communications sector mainly uses physical quantity measures, whereas the education and health sector uses a mixture of deflated wage bills and numbers of employees. Further details are available in CSO (1981). The only exception is that output in the public administration sector is based solely on the number of employees in the sector and excludes all HM Forces. One of the consequences of using such different output measures is that comparison of output growth between sectors is made very difficult. This applies not only to comparisons between individual service sectors but also to comparisons between manufacturing and service sectors. This is of particular importance when it comes to a discussion of the role of comparative levels of labour productivity in the relative growth of the

Table 9.2 Output growth in the service sector

Sector	Output growth rates (% p.a.)			
	1953–63	1963–73	1973–83	1953–83
Network services				
Utilities	5.3	4.8	1.5	3.9
Transport	1.8	3.4	−0.3	1.6
Communications	3.3	5.5	2.9	3.9
Private commercial services				
Financial and business	3.4	3.4	3.8	3.5
Distribution	3.3	2.6	−0.2	1.9
Other private	1.4	1.4	2.8	1.9
Construction	3.2	2.2	−2.1	1.1
Public social services				
Education and health	2.7	2.9	2.6	2.7
Public administration	0.2	1.8	−0.4	0.5
Manufacturing	3.0	3.5	−1.7	1.6

service and manufacturing sectors. This will be dealt with more fully in a later section of the chapter.

Output growth rates for each service sector using an index number of output at constant factor cost are shown in table 9.1. No simple, uniform growth patterns over the whole period are immediately obvious, as in the case of employment growth.

The fifties was a period of high output growth relative to manufacturing in financial and business services, distribution, education and health and construction, all labour intensive sectors experiencing high employment growth. There was also high output growth in utilities and communications, although employment growth (see table 9.1) was small in these capital intensive sectors. In the public administration sector, output based on an employment measure remained approximately static over the fifties, though again a slight rise is probably hidden by changes in SIC classifications. In fact, no sectors suffered output decline in the fifties, though slow output growth in the capital intensive transport sector was accompanied by employment loss. On the other hand, modest output growth in the labour intensive 'other private' services sector was accompanied by modest employment gain.

The next decade did not see a continuation of these trends in all sectors. High output and employment growth rates were maintained in the financial and business service sector and the public service education and health sector, both labour intensive sectors, and high output growth was accompanied by low or declining employment growth rates in the capital intensive communications, utilities and transport sectors, but the pattern differed in the remaining sectors. By definition, modest employment growth was accompanied by modest output growth in public administration, whilst 'other private' services actually saw output growth rise as employment fell, partly a consequence of the decline of domestic service. Similarly, construction and distribution almost maintained their high output growth rates of the fifties even though employment actually started to shrink.

The most noticeable feature of developments in the seventies was the drop in output growth rates in every sector with the exception of financial, business and 'other private' services. This paralleled a marked output decline in manufacturing over the same period. However, even though some sectors suffered decline in output, only one sector (construction) fared worse than manufacturing, i.e. all but one had higher relative output growth rates than manufacturing and all had higher employment growth rates.

Within this overall pattern of growth, many of the features of the earlier decades continued to show themselves at the sectoral level in the seventies. In particular, the capital intensive communications, utilities, and transport sectors, i.e. the network sectors, continued to differentiate themselves by their 'high output growth/employment decline' behaviour. Similarly, the labour intensive financial and education and health sectors continued to exhibit 'high output growth/high employment growth' characteristics. Amongst the remaining sectors, the high output growth experienced in the sixties in both construction and distribution collapsed, though the employment growth rate improved in the seventies in distribution as the effects of the self-service revolution worked themselves out. Finally, whilst public administration remained comparatively static in terms of output and employment growth, 'other private' services improved both its output and employment growth rates in the seventies.

Figure 9.1 summarizes overall developments over the last three decades. From this it emerges that there are three distinct clusters: two which exhibited consistent growth patterns over the whole period, and a third, remaining group which behaved in a more volatile fashion. Service sector growth, therefore, whether described in employment or output terms, has not developed evenly either over time or across sectors. This means that estimates of future employment growth and related policies must take account of the very different characteristics of each group.

Theories of Service Sector Growth

The growth of service sectors in industrialized economies this century has prompted both theoretical and empirical investigations into its causes,

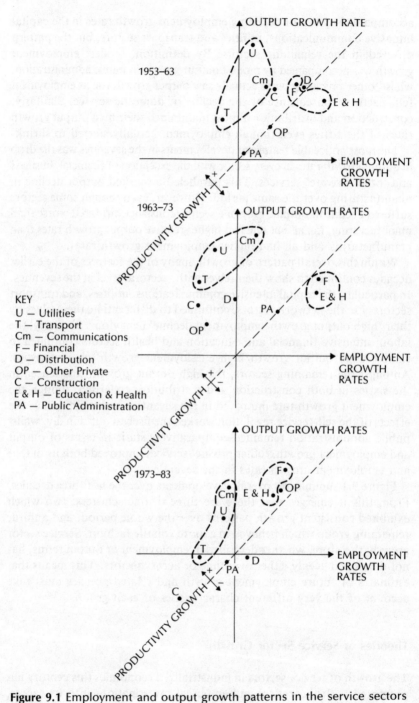

Figure 9.1 Employment and output growth patterns in the service sectors

though little consensus has emerged. The heterogeneous nature of service activities and the difficulties associated with defining and measuring commensurable growth rates for service sectors have played a large part in this. There has, however, been no shortage of competing explanations. In the following pages an attempt is made to develop a framework in which some of these mechanisms of service growth are first ordered and then discussed.

First, on the demand side, changes occur when the customer markets served alter. Changes can either take the form of consumption per capita increases, or of the net expansion, contraction or substitution of markets. Demographic movements and income elasticities are important demand side inducements to customer changes which have to be scrutinized in any investigation into the causes of service sector growth. Income elasticities are also important inducements to functional changes. These are closely related to customer changes but can be differentiated. They occur when the distribution of functional needs of a customer market alters without the customer market necessarily changing in any other way. As society as a whole gets richer, its ability to meet an expanding range of desires becomes greater.

Studies of income elasticity of demand (Fuchs, 1968; Gershuny and Miles, 1984) show that whilst income elasticities of demand cannot be sustained as an explanation of general growth in service activities relative to the demand for goods, income elasticity of demand is an important determinant of growth for particular services, such as tourism, or catering, or particular goods, such as consumer durables. Our sector studies of 'other private' services (Ed. Smith, 1985) suggested that there is likely to be substantial growth of demand in these sectors in the next 10 years. This means that the acceleration of output (and employment) growth in the seventies (tables 9.1 and 9.2) is likely to continue for the 'other private' services category and for hotels and catering within distribution.

Secondly, over the last 30 years there has been a growing demand for producer or intermediate services within other branches of the economy, and this phenomenon has often been put forward as a major contributor to the overall growth of service sectors. One way of looking at the growth of these services and the demand for them is to consider them under the heading of industrial structure changes. These occur when the provision of services by one industrial sector modifies the growth rates of other sectors providing either the same or competing services. International competition in traded services is an obvious case. One nation's airlines compete with another's and affect the growth rates of transport sectors. Another form occurs when a service originally undertaken within one sector, perhaps a manufacturing sector, is contracted out to another sector, a service sector. This is one mechanism whereby intermediate producer services expand. Yet another form of industrial change acting in the opposite direction occurs when the occupational structure of a manufacturing sector alters to accommodate more workers undertaking service tasks, as for example 'in-house' transport services.

The growth of producer services in service sector growth cannot be denied, but its relative importance has been the subject of debate. In Europe over the sixties, Gershuny and Miles (1983) maintain that whilst producer services were responsible for a considerable proportion of the net growth of services, they cannot be seen as the main source of service growth, social services accounting for more new jobs than producer services.

The employment growth data presented in table 9.1 can be used to examine the growth of producer services in the UK economy. The Browning–Singelmann (1978) definition of producer services corresponds very closely to our financial and business services category. This sector constituted only about 7 per cent of service sector employment as defined in this chapter in 1953, but by 1983 this figure had grown to over 12 per cent. Growth rates in this sector were consistently high over the whole period, but employment growth was at one time even faster in education and health.

Although by no means inconsiderable, the growth of intermediate demand could perhaps have been larger if it were not for the fact that increased demand for professional, technical, clerical and other specialized service occupations also grew within the manufacturing sector (Gershuny and Miles, 1983). Thus, although some service tasks were transferred from the secondary to the producer sectors of the tertiary economy, this transferral was 'retarded' by the greater growth of service occupations within the secondary sector.

Thirdly, service sector growth can be stimulated or retarded in various ways by changes occurring in the actual services provided. These encompass variations in the price of services, in the quality of services and in the range of services available. New modes of production can open the door to entirely new services via service product innovation and the extension of the range of services available. Services changes, therefore, are supply side changes heavily influenced by technological developments. Our sector studies suggested that new producer services associated with information technology will continue to expand very rapidly, for example in the provision of data banks.

The extension of service consumption as a mechanism of service sector growth has again been the subject of much debate. Many authors have argued that as societies get richer they consume more services and proportionately less goods (Clark, 1940; Fisher, 1939; Bell, 1974). However, Gershuny (1978) argues that in opposition to these trends there has been a shift to what he terms the self-service economy. In such an economy, services are actually replaced by the output of manufacturing sectors, e.g. laundries are replaced by washing machines at home, buses are replaced by cars, etc. In terms of the framework used so far in this chapter, shifts towards a self-service economy are a reflection of the changing relationship between service products and service functions, i.e. as time progresses, some functions are fulfilled by different products, be they new services or manufactured goods. Gershuny terms this type of change a 'modal shift'. In essence, self-service modal shifts can be seen as technologically induced supply side service changes acting in opposition to demand side income elasticity forces. Their undeniable existence in some service areas acts as a check to overall service sector growth.

Service changes which take the form of price and quality changes are often the result of labour productivity changes. For example, higher levels of labour productivity can lower costs and be reflected in lower end prices, increased demand and employment. Conversely, slow rates of labour productivity growth in services can raise prices, dampen demand and could lead to employment loss. However, a countervailing tendency could be for low service sector productivity growth, relative to manufacturing, to lead to a relative increase in the proportion of total employment taken up by the service sector. This is often put forward as a major source of relative service sector employment growth (Fuchs, 1968; Kuznets, 1972; Gershuny and Miles, 1983), although there are dissenting opinions (Marquand, 1978; Barras, 1982) which for the most part hinge on alternative measures of labour productivity in the service sectors.

Comparisons of labour productivity levels and growth rates in different sectors, particularly between manufacturing and services, are bedevilled by problems of mensuration. On the one hand, finding commensurate measures of the output numerator in different sectors is not made easy by the range of service outputs to be found, the marketed and non-marketed nature of these outputs and the difficulty of measuring improvements in the quality of outputs. On the employment denominator side, different patterns of full-time, part-time and self-employment complicate matters further.

Using the index numbers of output described in an earlier section of this chapter, labour productivity levels and growth rates over the last 30 years appear lower in services than in manufacturing, with the exception of the capital intensive network industries (table 9.3). Barras (1983) argues that growth rates in services and manufacturing over the period 1960–79 were almost identical, with public services making a lesser but still significant contribution. However, table 9.3 reveals that for the service sectors defined in this chapter, with the important exception of the network services, labour productivity growth rates over the last decade were

Table 9.3 Productivity growth in the service sector

	Productivity growth rates (% p.a.)			
Sector	1953–63	1963–73	1973–83	1953–83
Network services				
Utilities	4.7	5.9	2.0	4.2
Transport	2.5	5.2	1.4	3.0
Communications	2.0	4.7	3.2	3.3
Private commercial services				
Financial and business	0.6	−0.5	1.3	0.5
Distribution	0.2	2.6	−0.9	0.7
Other private	−0.1	2.8	0.6	1.1
Construction	1.5	4.0	0.5	2.0
Public social services				
Education and health	−1.4	−1.0	0.8	−0.6
Public administration	0.0	0.0	0.0	0.0
Manufacturing	2.4	4.0	1.8	2.7

generally lower than manufacturing growth rates. Substituting full-time equivalent employment figures in the denominator undoubtedly raises the productivity performance of some services (see for example Smith, 1986a, p. 23) but the weight of available evidence still tends to suggest that, in the labour intensive service sectors at least, labour productivity growth rates have not approached those to be found in the manufacturing sector. Slower productivity growth has thus almost certainly 'contributed' to the employment growth of services relative to manufacturing, though mensuration difficulties continue to obscure to what extent.

Our own sector studies suggest that although information technology is of steadily growing importance in all service sectors, its greatest impact on labour productivity will be in financial services and network services. Its impact on labour productivity in other private services and public social services, in so far as it can be measured, is likely to be rather small, at least for the next few years, so that steady growth in demand in these sectors is likely to have positive employment generating effects.

In the financial and business service sector the rapid growth in capital intensity associated with computerization in banking and insurance (de la Mothe, 1986) and the associated increase in labour productivity has led to a slowing of employment growth in this subsector, despite continuing high output growth. Nevertheless, there is a wide area of business services (e.g. software consultancy, plant hire, and new financial services) which seem likely to expand very rapidly since they are associated with the provision and expansion of new types of service, rather than capital-deepening in established modes of service provision.

To sum up the discussion in this section: no single theory can explain the overall growth in service sector employment from 1953 to 1983 relative to manufacturing employment. It must be attributed, as in other OECD countries, to a combination of the following factors:

(1) The growth of public social services and administration, unaccompanied by any significant change in measured labour productivity. The pressures of consumer 'demand', although undoubtedly strong, were in this case expressed indirectly through the political system and budget process. Public policies are decisive in influencing the trend of employment in these services, but experience in the United States shows that private market demand for education and health is also very high. One Annium estimate was that over 3 million new jobs would be created in education and health by 1995 (Persnick, 1984).

(2) An increase in demand for intermediate (producer) services, especially financial and business services, based in part on new services, such as computer software, and also on a shift of some service activities from 'in-house' to a 'contracted out' basis of provision. The distinction between 'goods' and 'services' is in any case becoming blurred in this area as more firms supply both 'hardware' and 'software'.

(3) The growth of some private services to final consumers, characterized by high income elasticities of demand, such as catering and tourism and other private services, offsetting declining demand for those private services which were shifting to a 'self-service' mode of provision through technical change in consumer durables (laundries, cinemas, domestic service).

In all of these three areas, the growth of output and demand outstripped the (relatively slow) rate of labour productivity growth. Taken together, these expansionary forces more than compensated for the declines of employment in the network services and construction, sectors which (like manufacturing) experienced rather high growth of labour productivity, based on capital-embodied technical change. The construction sector was also heavily affected by cyclical factors and public policy changes.

Despite the manifest difficulties in constructing satisfactory output and productivity measures, this discussion has brought out some important distinctions between the various service sectors, which relate in turn to capital intensity and changes in the pattern of investment. It is to these characteristics that we now turn. It will then be possible to assess the overall potential for future service employment growth in relation to manufacturing.

The Changing Capital Stock

The service sector trends in employment, output and labour productivity delineated so far have been accompanied by changes in the structure of the capital stock of each sector. This section and the ones which follow address these changes. In particular, emphasis is placed on changes in that part of the capital stock classified as plant and machinery, i.e. on the component of capital stock most likely to reflect changes in the technologies utilized within each sector.

The first point to note about the structure of the capital stock in the service sector is that the share taken up by plant and machinery is far less than in the manufacturing sector. In 1983 only 30 per cent of gross capital stock in the service sector as defined in this chapter consisted of plant and machinery, the remainder consisting primarily of building stock. By contrast, plant and machinery accounted for nearly two thirds of

Table 9.4 Investment in plant and machines in the service sector

Sector	Share of service sector investment in plant and machinery (%)			
	1953	1963	1973	1983
Network services	85.5	81.2	62.3	40.0
Utilities	35.7	43.5	16.6	15.9
Transport	37.2	23.2	26.4	12.3
Communications	12.6	10.5	19.3	11.8
Private commercial services	8.3	13.7	26.9	51.0
Financial and business	1.2	2.6	11.3	31.5
Distribution	5.5	8.7	12.3	14.3
Other private	1.6	2.4	3.4	5.2
Construction	3.4	4.2	3.9	2.0
Public social services	2.8	4.8	6.9	7.1
Education and health	1.9	4.0	5.0	3.9
Public administration	0.9	0.9	1.9	3.1
All services	100.0	100.0	100.0	100.0

manufacturing capital. It should also be noted that although the proportion of total capital stock taken up by plant and machinery in the service sector as a whole has grown over the last three decades, this growth has been slow despite a relatively high absolute growth in plant and machinery capital stock. However, the share of plant and machinery in new investment in the service sector in 1983 had risen to 50 per cent.

Table 9.4 provides some figures on the relative shares of each service sector in total service sector plant and machinery. Not unexpectedly, utilities, transport and communications dominate the scene, vehicles being classified as part of the plant and machinery total in the case of the transport sector. However, table 9.4 also reveals that the proportion taken up by these sectors has fallen sharply as a result of: (1) increases in the stock of plant and machinery in every other sector, the financial sector in particular; and (2) the dramatic fall in the share of the transport sector due to the decimation of the UK merchant fleet, the number of vessels halving in the seventies alone (Brodie, 1984). The actual rates of growth of plant and machinery stock are given in table 9.5, from which it is obvious that the highest rates of growth in each decade occurred in those sectors with the smallest plant and machinery base. In fact, rates of growth in all sectors other than the network sectors were higher than in manufacturing in each of the three decades considered.

Table 9.5 Plant and machinery capital stock growth in the service sector

Sector	Plant and machinery growth rates (% p.a.)			
	1953–63	1963–73	1973–83	1953–83
Network services	2.9	4.3	0.4	2.6
Utilities	4.7	5.2	0.4	3.4
Transport	1.6	1.7	−1.9	0.5
Communications	2.1	8.4	3.9	4.8
Private commercial services	5.4	9.3	9.3	8.0
Financial and business	6.0	14.8	14.9	11.9
Distribution	5.3	7.7	6.0	6.3
Other private	5.6	8.4	6.3	6.8
Construction	9.0	8.0	2.9	6.6
Public social services	13.3	10.4	5.6	9.8
Education and health	14.6	10.4	4.5	9.9
Public administration	10.1	10.5	8.3	9.6
All services	3.6	5.4	3.0	4.0
Manufacturing	4.4	4.1	2.2	3.6

Another important point to note about current investment is that in absolute terms expenditure on plant and machinery in the service sector is much greater than in the manufacturing sector, nearly three times as much in fact, but this is mainly due to the extremely high capital intensity of the network services and the role of the financial service sector in the hiring out of plant and machinery to other industrial sectors. Nevertheless, total investment in plant and machinery in the lowly capitalized non-network service sectors still currently exceeds that in manufacturing.

Investment growth rates in services in the seventies fell sharply compared to those of the sixties. In construction, transport, communications and education and health, investment actually declined. Even the dynamic financial services sector could not maintain the peaks of the sixties, though growth remained very high and was spectacular in subsectors such as business services as a result of the combined influence of the growth of plant-hire (especially computers) and the computerization of banking and insurance.

Two points need to be emphasized about this investment slowdown in the seventies. On the one hand, it was accompanied by a general fall in service sector labour productivity growth rates. This provides further *prima facie* evidence for a link between investment in plant and machinery and labour productivity growth. On the other hand, in most sectors investment growth rates remained higher than in manufacturing. What investment took place in plant and machinery in the seventies and early eighties took place primarily in the service sectors rather than in manufacturing. (However, the rapid increase in plant hire from the service to the manu-facturing sector must be taken into account.)

The amount of capital employed per person in the service sector as a whole is roughly comparable with the figure for manufacturing: around £35,000 in 1983 at 1980 prices. However, this average is heavily influenced by the network services. Moreover, since the proportion of total capital taken up by plant and machinery is smaller in the service sector, the value of the plant and machinery employed per person is only about 50 per cent of the manufacturing figure, i.e. around £11,000 compared to £22,000. For the rest of this section, whenever the term capital intensity is used, the ratio implied is the restricted one of value of plant and machinery per employee.

It is obvious from table 9.6 that the network sectors are the most truly capital intensive sectors, all the rest having very low capital intensities in comparison. It is true to say that the latter have gone some way towards 'closing the gap'. The least capital intensive sectors experienced the highest growth rates in capital intensity over the whole of the last three decades as investment in plant and machinery increased, but the gap in the early eighties was still very considerable.

Table 9.6 Plant and machinery capital intensity in the service sectors

Sector (all services in 1953 = 100)	Plant and machinery capital intensity indices				1983 £'000 per employee
	1953	1963	1973	1983	
Network services	442	601	1026	1194	
Utilities	909	1368	2556	2800	
Transport	343	435	616	600	
Communications	329	356	760	1150	
Private commercial services	22	29	71	158	
Financial and business	21	29	85	295	
Distribution	24	30	64	110	
Other private	19	29	78	118	
Construction	16	32	86	150	
Public social services	5	14	28	45	
Education and health	6	18	34	44	
Public administration	3	8	19	46	
Manufacturing	113	165	264	467	22
All services	100	117	186	237	11

Conclusions

Taking into account the evidence of our sector studies relating to the diffusion of new technology, it is likely that substantial differences will remain between the pattern of investment and employment growth in those tertiary sectors which resemble manufacturing in their capital intensity and capital embodied technical change, and other service sectors, especially public social services which are still extremely labour intensive.

This means that the methods used in chapters 6 and 7 to estimate possible future employment trends, based on a capital vintage model, are generally inappropriate for trends in most service sectors, even if better measures of output and productivity were available. Our assessment of future employment trends in most service industries are therefore based to a much greater extent on a qualitative assessment derived from our sector studies, and on the consideration of possible policy changes.

The overall conclusions from this chapter and from these sector studies are as follows:

(1) It is essential to distinguish between different service sectors: they differ very greatly in their characteristics even though they are all being affected by growth in capital intensity.

(2) When four major categories are distinguished, then it emerges that the 'network services' are the only ones which have resembled manufacturing in their pattern of output and employment growth. What has been said about the manufacturing sectors in chapter 8 applies in large measures to the 'network services' as well. However, just as the analysis in chapter 8 distinguished some parts of manufacturing (especially computers and electronics) where there is still some prospect for employment growth, this applies also to the communication subsector. Not surprisingly, these sectors form the 'core' of the emerging IT sector discussed in the next chapter.

(3) When we turn to the private commercial service sector, the pattern of growth of output and employment is quite different. Despite the general growth of plant-and-machinery capital intensity, the levels remain much lower than in manufacturing or network services, except in a few specialized subsectors. The private services account for nearly a third of total UK employment, i.e. much more than manufacturing, and this share is likely to rise as both producer services and income elastic consumer services will benefit from future economic growth.

(4) Trends in public social services and construction employment are dominated by public policy. The implications of this situation are dealt with in chapters 11–13.

10

The Newly Emerging Information Technology Sector

Luc Soete

Despite a vast and rapidly increasing literature on the subject there is no agreed common definition of what exactly the 'information technology' (IT) sector should or could comprise. Worse, there is not even an agreed common methodological framework within which the IT concept could be studied. Some studies opt for an occupation-based definition, attempting in the first instance an inventorization of 'information-related' activities; others are more of the traditional sectoral type, disagreeing more often than not as to which sectors should or should not be included. Some of the most recent studies limit the information sector to the so-called 'tradeable' information sector, emphasizing the fact that there are a wide range of information-related activities which, particularly within the firm, will not be separately valued and will consequently not be included in any 'information sector' definition.

The UK Government's Information Technology Advisory Panel includes in the 'tradeable' information sector the 'supply of financial and business information, printing and publishing, on-line technical information, consultancies, . . . the entertainment industry and aspects of education and training services' (ITAP, 1983, p. 7). Such a definition is almost exclusively concerned with inventorizing the traditional information service sectors, and is at complete loggerheads with, e.g., NEDO's narrow, and predominantly manufacture-based IT definition (computer equipment and services, and telecommunications equipment). Clearly the manufactured inputs required to 'produce' information activities should be fully part of any definition of the 'tradeable' information sector. It is, as emphasized in chapter 8, the manufacturing sector which has been at the core of the technological transformation of most traditional information activities,

189

Table 10.1 Components of information labour force (percentage of economically active)

	Canada		France		Germany		Japan		United Kingdom		United States	
	1951	1971	1954	1975	1950	1982	1960	1975	1951	1981	1950	1980
Information producers	4.4	7.6	3.6	6.4	3.1	6.9	2.1	4.5	3.9	8.8	5.0	9.7
Scientific & technical	2.6	5.3	0.6	1.5		1.7	0.6	1.0	0.5	2.0	1.3	2.1
Consultative services			1.8	3.0		2.7	0.5	1.1	1.1	3.3	1.9	3.9
Information gatherers	0.8	1.1	0.3	0.5		0.3			1.0	1.3	0.3	0.4
Market search & co-ordination specialists	1.0	1.2	0.9	1.6		2.2	1.0	2.4	1.3	2.3	1.5	3.3
Information processors	20.1	25.2	13.4	19.7	11.2	10.3	12.3	20.6	18.3	24.1	21.2	28.6
Administrative & managerial	10.4	12.3	6.3	6.7		4.8	2.3	4.4	4.5	7.0	8.6	11.5
									3.2	3.6	1.4	4.0

	1	2	3	4	5	6	7	8	9	10
Process control & supervisory	9.7	1.5	3.2	1.1	10.0	16.2	10.6	13.5	11.2	13.1
Clerical & related	12.9	5.6	9.8	14.4						
Information distributors	2.5	1.9	3.9	3.0	1.9	2.4	2.0	4.3	2.3	4.4
Educators	2.1	1.6	3.7	2.9	1.8	2.2	1.7	3.9	2.1	4.2
Communication workers	0.4	0.3	0.2	0.1	0.1	0.2	0.3	0.4	0.2	0.2
Information infrastructure	2.3	1.4	2.1	4.6	1.6	2.1	2.5	3.6	2.2	3.1
Information machine workers	0.9	0.6	0.7	3.7	1.6	1.4	1.4	1.6	1.0	1.7
Postal & telecommunications	1.4	0.8	1.4	0.9	0.0	0.7	1.1	2.0	1.2	1.4
Total information	29.3	20.3	32.1	34.8	17.9	29.6	26.7	41.0	30.7	45.8

and has led to the emergence of a 'new' information technology sector, cutting across the manufacturing–service divide.

In a similar vein, one should also remain sceptical with regard to the attempts at developing occupational, rather than sectoral definitions of the information sector mentioned earlier on. Such definitions typically aim at identifying the predominantly information-based occupations, whether primarily information producer or information processing in nature, or more indirectly related to the servicing and maintenance of the information infrastructure. Attempts by the OECD (ICCP unit) at developing such internationally comparable statistics have by and large been unsuccessful, the existing national occupational data providing a too general and in many ways out-dated basis to allow for such information occupations inventory calculations. For what it is worth, table 10.1 provides some hint as to the percentage of the economically active engaged in various information occupations, on the basis of the most recently available OECD statistics.

As in the case of the UK Information Technology Advisory Panel, the occupational approach in identifying the information sector is too much based on the notion that it is the provision of information or the knowledge base of a particular activity, whether this was as a lawyer in 1950 (or 1850) or as a computer systems analyst in the 1980s, which is of relevance in the study and analysis of the newly emerging information technology sector. In Machlup's pioneering study: *The Production and Distribution of Knowledge* (1962), 42.8 per cent of the economically active were already identified in 1959 as belonging to 'knowledge-producing' occupations.

In this chapter we will analyse the information technology sector from a traditional sectoral point of view. This is the most straightforward and from a statistical point of view most easily implementable framework, popular with most empirical studies on the subject. At the core of the IT sector one finds in this definition the IT producing sectors, both manufacturing and services. Table 10.2, based on a recent Office of Technology Assessment report (OTA, 1985) on the US IT industry, illustrates, e.g., the growth in employment in the sector, based on such a definition. McKinsey, in their IT report for the EEC, define the sector somewhat more broadly, also including, as illustrated in table 10.3, robotics, avionics, etc. Clearly in this latter definition a number of predominantly IT-using sectors are also included in the IT sector. Here we will focus in the first instance on the IT producing sector. Some predominantly IT using sectors, such as some of the manufacturing sectors listed in table 10.3 or an increasing number of the 'tradeable' information service sectors mentioned earlier on, will however also be considered as forming fully part of the IT sector. As we will see in the next section, apart from constantly broadening this group of sectors, IT in its sectoral user impact will however have a significantly different impact on the

Table 10.2 Employment levels in the US Information Technology Industries'
Employees (in thousands)

	1972	1982	Percentage change 1972–82
Manufacturing[a]			
Computers	145	351	+ 142
Office equipment	34	51	+ 50
Radio & television receiving sets	87	63	− 28
Telephone & telegraph equipment	134	146	+ 9
Radio & television communications equipment	319	454	+ 42
Electronic components	336	528	+ 57
Totals, manufacturing	1055	1593	+ 538
Services			
Telephone & telegraph	949	1131	+ 11
Computing[b]	149	360	+ 141
Radio & television broadcast[c]	68	81	+ 19
Cable television[d]	40	52	+ 30
Totals, services	1206	1624	+ 618

[a] Estimates provided by the US Dept of Commerce, Bureau of Industrial Economics
[b] Figures are for 1974 and 1983. *Source:* US Industrial Outlook, 1984
[c] Figures are for 1979 and 1983. *Source:* Federal Communication Commission in telephone interview with OTA staff, May 1984
[d] Figures are for 1981 and 1982. Ibid. (FCC)
Source: OTA, 1985

IT-using service sectors as compared to the IT-using manufacturing sectors.

The Economics of Information Technology

The main argument we would like to put forward in this chapter is that information technology is likely to have a diametrically opposite impact on the service sector of the economy as compared to the manufacturing side. The reason for this has to do with the nature of the new technology and its capacity to store, process and disseminate information at minimal costs. In the case of services, it will be argued, this will lead to the increased 'tradeability' of a number of service activities, whereas in the case of manufacturing, the opposite might well be true. We start the analysis with the service sector.

Table 10.3 The IT sectors

Sector	Description	Product examples
1 Industrial automation	Equipment used in specific factory and design automation applications	*Numerical controllers *Process control *Robotics
2 Office automation	Equipment used in office automation applications	*Word processors *Executive workstations
3 Telecommunications	Voice and data-communication equipment currently handled by PTTs	*Central office switching *Modems
4 Defence, aerospace and communication	Equipment used in defence, aerospace and communication	*Avionics *Guidance systems *Radar
5 Other industrial	Miscellaneous equipment sold to industrial users and equipment incorporated in other products	*Medical electronics *Automotive electronics
6 Consumer	Consumer electronics	*Video equipment *Home information systems
7 Computers	General purpose processors and peripherals and complex data processing systems	*Mainframe computers *Minicomputers *Terminals
8 Components	Components used in the manufacture of products in all other sectors	*Integrated circuits *Discrete semi-conductors *Lasers
9 Software and services	Software and computer services sold independently of hardware	*Remote computer services *Packaged software

Source: McKinsey

The Impact of IT on Services

Services will be defined here, following Quinn (1986), as those activities (sectors) where output is essentially consumed when produced. This might well be considered as a rather narrow definition and, as chapter 9 has illustrated, one which would only cover a limited number of sectors presently understood to be service sectors. It is however an analytically useful definition and one which allows for a far more clear-cut interpretation of what the potential impact of IT might be on services. In parenthesis it is worth noting that it also provides an intellectual argument as to why economists have generally tended to ignore the study and analysis of service activities.

Typically, information technology, practically by definition, will allow for the increased tradeability of service activities, particularly those which have been most constrained by the geographical or time proximity of production and consumption. By bringing in a space or time/storage dimension, information technology will make possible the separation of production from consumption in an increasing number of such activities.

This was certainly the case with regard to the invention of printing in the Middle Ages and the impact this first new information technology had on the limited tradeable 'service' activity of monks copying manuscripts by hand. It was the time/storage dimension of the new printing technology which opened up in the most dramatic and pervasive way access to information, and led, to use Marx's words, to the 'renaissance of science', the growth of universities, education, libraries, the spreading of culture, etc. This opening up, 'tradeability' effect would become of far more importance to the future growth and development of Western society than the emergence of a new, in this case purely manufacture-based, printing industry. Similar less pervasive, but nevertheless crucial 'trade opening-up' impacts could be ascribed to some of the more modern 'copying' inventions, be it of written text, spoken language, music or image (photocopying, recording, taping), allowing, amongst others, for the emergence and growth of 'mass' entertainment, as opposed to the space and time constrained individualized entertainment.

In the case of the invention of the telephone it could be said that it was in the first instance the capacity of the new technology to bridge geographical space which opened up whole new tradeable markets. In this case this led both to the emergence of a new, predominantly service sector, maintaining and servicing an increasingly extensive and international infrastructure network and a new rapidly growing manufacturing sector. Again though, the indirect 'tradeability' effects in terms of the use of telecommunications in an increasing number of service activities and the opening up of such activities to private market valuation were significant, including the international trade effects.

In the case of the current 'new' information technologies and their potential to collect, store, process and diffuse enormous quantities of information at minimal costs, both the time/storage and space dimensions of the new technology can be expected to bring about the further opening up of many service activities, increasing their domestic and international tradeability. As in the case of the telephone it is likely that the 'new' emerging IT manufacturing sector (in the first instance the computer manufacturing sector) will remain relatively small compared to the growth and size of the new IT service producing sector. Again, though, it will be the indirect 'trade' effects resulting from the use of IT in many services, and in particular in those IT-using services listed above as forming part of the UK 'tradeable' information sector, which will be most important both in terms of employment and output growth. Probably one of the most important features with regard to these indirect 'tradeability' growth effects will be with respect to international trade and the United Kingdom's likely future balance of payments constraint – a point to which we come back below.

The Impact of IT on Manufacturing

With regard to manufacturing, the impact of IT could well be characterized as exactly of the opposite kind. Rather than bringing time/storage or space between production and consumption as in services, IT in its manufacturing user impact will in the first instance aim at reducing the time/storage or space dimension between production and consumption.

As brought out in chapter 4, many of the most distinctive characteristics of the 'new information technology paradigm' are directly related to the potential of the new technology to link up networks of component and material suppliers, thus allowing for reductions in storage and production time costs – typified in the so-called just-in-time production system (Schonberger, 1982). At the same time the increased flexibility associated with the new technology allows also for a closer integration of production with demand, reducing the firm's own storage and inventory costs – what could be typified as just-in-time selling (see e.g. Benetton). Both features clearly work in the opposite direction from what was said above with regard to services, i.e. they aim at reducing the time/storage dimension between production and consumption. However, in doing so they will also reduce the 'tradeability' of a number of those intermediary, storage and inventory manufacturing activities.

The increased potential for flexibility and decentralization generally associated with the new information technologies (Perez, 1985) can also be expected to reduce the geographical space dimension between production and consumption in many manufacturing sectors. The relative increase in physical (production or person) as opposed to information transport costs might well lead to closer location of production units to consumer markets. Again and in this case more with respect to the international trade implications, it could be said that in contrast to services, IT could well reduce the international tradeability of many manufactured commodities. This is already, but not necessarily for the reasons set out here, the case with regard to the motives underlying much foreign investment in developed countries.

The Impact of IT on International Trade

The long-term implications of these rather different 'trade' impacts of IT on services and manufacturing for the future balance of payments constraint of a country such as the United Kingdom are far-reaching and have so far received little attention. As was mentioned in chapter 8, the 'manufacturing' sector balance of payment constraint will over the next decade increasingly emerge as one of the major bottlenecks to a sustained period of more rapid recovery growth in the United Kingdom.

Leaving aside transport producer services, where the 'trade' impact of IT is likely to resemble more the arguments set out above with regard to manufacturing, i.e. acting more as a substitute for the physical transport of goods or persons, than as a 'trade opening-up' complement, the trade surplus in services reached £6000 million in 1985, compared to only £2000 million in 1983 and £800 million in 1981. At the same time the trade deficit in manufacturing (excluding oil) amounted to some £4000 million in 1985, compared to a trade surplus of some £3300 million in 1981. In so far as services, as a result of the further use of IT, will increasingly become internationally 'tradeable' and contribute to the trade balance with the United Kingdom enjoying both an absolute (language and institutional – the City, time zone) and comparative advantage in many such IT-using services, traditional balance of payments growth constraint views might well prove to represent a too pessimistic and too static picture of the United Kingdom's likely future growth prospects. This will, however, depend on the extent and speed of the application of new technology in the service and manufacturing sectors, as is already evident, for example, in the rapidly increasing Japanese competition in financial services.

In other words, claims that growth in its international dimension 'still depends primarily on manufacturing' (House of Lords, HMSO, 1985) or that whatever services are tradeable, 'trade in services is subject to the same influences as trade in goods' (Bank of England, 1985), are likely to be too restrictive with regard to the impact IT might have on the tradeability of services and their increasing contribution to future economic growth.

The extent to which the UK's trade advantage in some of these services is an intrinsic structural feature of the UK economy remains, however, open to question and needs to be cautioned. As in manufacturing, the UK 'core' IT-producing service sectors have been confronted with an increasing trade deficit over the last couple of years. As table 10.4 illustrates, it is in the first instance some of the more traditional, now predominantly IT-using services which have continued to enjoy an increasing trade surplus. Yet, despite having done so the UK's share of world exports in these services has declined steadily from 12 per cent in 1969 to 8.2 per cent in 1985. Thus, whereas IT is likely to increase further the proportion of services that are potentially 'tradeable', it will do so also with regard to the United Kingdom's main competitors, possibly

Table 10.4 UK receipts from services (£ millions, 1983)

	Gross	Net
Sea transport	3054	− 884
Civil aviation	2665	428
Travel	3655	− 399
Insurance[a]	1606	1595
Banking[a]	4571	1842
Other financial[a]	1975	1941
Telecom., postal, film, TV	643	− 14
Royalties, etc.	1292	183
Construction & consultancy[b]	1409	1409
Other services	3253	1297

[a] These include an element of interest, profits and dividends in addition to service activities.
[b] The debit figure for construction and consultancy services is small and included in the 'other services' net figure.
Source: British Overseas Trade Board as quoted in Mansell, 1986

eroding the United Kingdom's absolute advantage in some of the more traditional services activities, where institutional rather than international competitiveness factors have been the main reason behind the country's trade surplus.

The arguments advanced here remain however speculative. We return to them in the conclusions.

The UK 'IT Sector': A Guesstimate

The implications of the previous analysis for output and employment growth in the various components of the IT sector are relatively straight-forward. Independent of any price or demand-induced compensation effects one would expect significantly different patterns to emerge, particularly with regard to employment growth, in the manufacturing IT-using sectors and the service IT-using sectors. In the former sector the arguments advanced earlier would suggest that the gains in productivity achieved through the use of IT will have been accompanied by little new 'trade', if anything rather the opposite; employment is therefore likely to have fallen. In the IT-using services by contrast, one would expect to witness the biggest output gains in new tradeable areas, in all likelihood accompanied by employment growth. With regard to the IT-producing sector, the distinction between manufacturing and services is clearly of less importance. One would expect rapid output growth in both sectors accompanied with employment gains depending on the underlying rate of productivity growth.

To check for these differential growth trends is no easy undertaking. Output data with regard to the service sector are, as illustrated in chapter 9, a statistician's invention, based in many cases on employment trends. Employment data, on the other hand, exist on a disaggregated common basis only since December 1981 as already mentioned in chapter 8. Finally, whereas a definition of the IT-producing sector is always feasible, it is a far more hazardous undertaking to select the 'predominantly' IT-using sector.

It should also be stressed that to assume such output or employment trends would be purely IT-induced and would be independent of any price-induced output or employment compensation effects is rather heroic. However, over a relatively short period such as the one imposed here, i.e. the period since 1981, which is also from a macroeconomic demand compensation point of view a rather depressing period, such an assumption might appear less heroic.

Bearing in mind what was said in the introduction, we have ourselves been guided in the selection of the predominantly IT-using sector by the UK IT advisory panel's list of information sectors. Table 10.5 provides information on the 1985 employment level and trend in employment growth/decline over the last 5 years in the various components of the IT sector, separating out the IT-producing sector from the 'predominantly' IT-using sector. The IT-producing sector comprises on the manufacturing side the computer, electronics and telecommunications equipment industries and on the service side the telecommunications and computer services sectors. This is a relatively narrow definition of what the IT-producing sector amounts to. It comprises however the 'core' IT manufacturing industries, responsible for the emergence of the new information technologies and the 'core' IT service sectors essential for the application, maintenance and service of these hardware technologies.

Taken together, these IT-producing sectors represented only 3.3 per cent of total employment in the UK. Furthermore, in terms of employment growth, the total increase in employment in the IT-producing sectors over the last 5 years was a mere 500. Most of this growth occurred in the manufacturing IT-producing sector (+6000) particularly in the non-telecommunications part (+19,600). On the services IT-producing side, the decline in telecommunications clearly outweighed the gains in computer services. The decline in employment in telecommunications, both manufacturing and services, illustrates the predominance of the productivity gains associated with the replacement of electromechanical switching equipment by electronic digital equipment over new markets. The contrast between the relatively slow growth in output in telecommunications despite

Table 10.5 Employment levels in the UK information technology industries (Great Britain)

		1981	1985	Growth/
		(1000s)		Decline
I IT-producing sector				
Manufacturing				
3302	Data processing equipment	72.2	75.3	3.1
3441	Telegraph & telephone apparatus and equipment	58.1	44.5	−13.6
3442	Active components	25.9	26.9	1.0
3443	Radio & electronic capital goods	87.5	94.6	7.1
3444	Components (passive)	29.0	32.9	3.9
345	Other electronic equipment	125.3	129.8	4.5
Total manufacturing		398.0	404.0	6.0
Services				
7902	Telecommunications	231.7	219.6	−12.1
8394	Computer services	56.0	62.6	6.6
Total services		287.7	282.2	−5.5
Total IT-producing		685.7	686.2	+0.5
% of all employees in employment (Great Britain)		3.2	3.3%	
II Predominantly IT-using sector				
Manufacturing				
3221	Metal working machine tools	38.8	31.5	−7.3
3289	Precision engineering	207.4	158.8	−48.6
3710	Measuring, precision instruments	58.7	65.0	6.3
475	Printing and publishing	351.3	348.8	−2.5
4930	Photo cinematographic processing	14.2	12.3	−1.9
Total manufacturing		670.4	616.4	−54.0
Services				
81	Banking	469.3	525.6	56.3
82	Insurance	225.1	246.7	21.6
83(−8394)	Other business services	798.5	918.6	120.1
94	Research and Development	119.8	134.5	14.7
9741	Radio and TV	69.3	72.8	3.5
Total services		1682.0	1898.2	+216.2
Total IT-using		2352.4	2514.6	+162.2
TOTAL		3038.1	3200.8	+162.7
% of all employees in employment (Great Britain)		14.4	15.3	

Source: *Employment Gazette*, August 1984, April 1986

virtual total protection from international competition and the significant growth potential resulting from the new technology is however striking. There is, we would argue, little doubt that this failure to develop new services and new 'trade' on a large scale was at least partly the result of the 'infrastructure monopoly' the telecommunications industry enjoyed for so long, associated with the oligopolistic equipment supply arrangements which prevailed.

On the IT-using side, the difference between manufacturing and services is, as expected, far more clearcut. Most of the predominantly IT-using manufacturing sectors, defined here – again in a relatively narrow fashion – as 'robotics', instruments (including precision engineering), and printing (including photocopying), witnessed significant employment losses over the last 5 years; in total some 54,000. By contrast, the predominantly IT-using service sectors, defined here as financial and business services, research and development, and radio and TV, saw their employment growth over this same period increase by some 216,000. The definition of 'predominantly' IT-using, particularly with regard to services, is of course highly subjective. We have included here most of the sectors suggested by the Government's Information Technology Advisory Panel as belonging to the 'tradeable' information sector. Some of these service sectors listed in table 10.5 include undoubtedly also activities which are not really at the 'core' of IT use in services. However, most of the sectors listed have witnessed a significant increase in tradeable activities as a direct result of the use of IT. This is certainly the case in banking where the further use and introduction of IT has allowed the gamut of services (including international service activities) offered by banks to increase significantly. Thus, despite significant productivity gains in the handling of money, employment increased over the last 5 years by some 56,300. In 'other business services' (excluding computer services, which was classified as IT-producing) this is likely to be even more the case, with plenty of opportunities for new demand outlets, often, as in the case of the small instant print and photocopy shops, the direct result of the use of IT.

The total increase in employment in the IT sector over the last 5 years, taking into account both the producer and user sectors, amounted to some 162,700. Primarily because of the inclusion of the predominantly IT-using service sectors, employment in IT represented in 1985 some 15.3 per cent of total UK employment; in 1981 only 14.4 per cent. This has to be set against the 3.3 per cent of the IT-producing sector.

The latter sector, as we illustrated in chapter 8 and discuss in more detail in the next section, represents though more than 30 per cent of total industrial R & D. In terms of output, table 10.6 provides some crude guesstimates as to the contribution of the IT sector, and its main components, to GDP

Table 10.6 The IT sector's share of GDP (in %): 1978–84

	IT-total	IT-producing			IT-using			Manuf. [excl. IT]	Services (incl. construction) [excl. IT]	Other (incl. energy)
		Total	Manuf.[a]	Services[b]	Total	Manuf.[c]	Services[d]			
1978	17.6	3.3	1.4	1.9	14.3	2.4	11.9	25.4	52.6	4.4
1979	18.2	3.5	1.5	2.0	14.7	2.4	12.3	24.5	51.9	5.4
1980	19.4	3.8	1.6	2.2	15.5	2.3	13.2	22.7	52.4	5.5
1981	20.2	3.9	1.6	2.3	16.3	2.3	14.0	21.6	51.1	7.1
1982	20.8	4.1	1.7	2.4	16.7	2.1	14.6	21.2	51.4	6.6
1983	21.3	4.4	1.9	2.5	16.9	2.0	14.9	21.1	50.9	6.7
1984	22.3	4.8	2.2	2.6	17.5	2.3	15.2	20.9	50.3	6.5
Average Annual Growth Rate 1978–84	4.95	7.33	8.68	6.26	4.35	−1.16	5.26	−2.26	0.30	5.82

[a] 1980 GDP weights: AH 330: 0.36; AH 3441: 0.279; AH 3443: 0.484; AH 3444: 0.195; AH 3452: 0.091; AH 3453: 0.202.
[b] 1980 GDP weights: AH 7902: 1.832; AH 8394: 0.376.
[c] 1980 GDP weights: AH 3221: 0.175; AH 3289: 0.339; AH 3710: 0.179; AH 3733: 0.042; AH 475: 1.605.
[d] 1980 GDP weights: AH 81,82,83 (excl. AH 8394): 10.650; AH 940: 2.270; AH 9741: 0.277.

over the period 1978–84. That contribution according to the estimate given in table 10.6 has increased from 17.6 per cent in 1978 to 22.3 per cent in 1984. The IT sector is now more important in terms of its contribution to GDP than the manufacturing sector. Even services (excluding IT-services) have seen their share of GDP decline from 52.6 per cent in 1978 to 50.3 per cent in 1984. In terms of output growth rates, figure 10.1 illustrates the rapid growth of the IT sector over the period considered. As indicated in table 10.6, this is primarily the result of the rapid output growth in the IT-producing sectors, and the increase in 'tradeable' output in IT-using services.

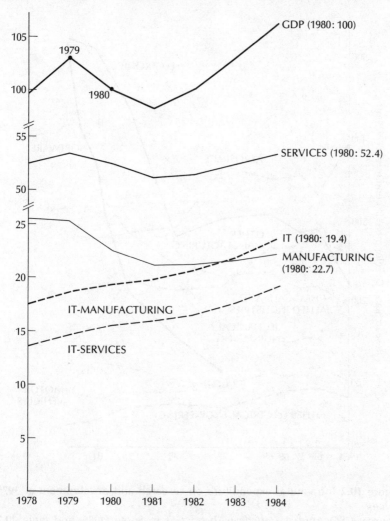

Figure 10.1 GDP growth by sector 1978–84 (GDP, 1980 = 100)

Information Technology: The Technological Evidence

From a purely technological perspective, the importance of the cluster of information technologies is well recognized today. We have already illustrated in chapter 8 that in terms of R & D expenditure, e.g. the manufacturing arm of the sector (electronics, telecommunications, computers) represented now near to 30 per cent of total UK manufacturing R & D. However, as figure 10.2 illustrates, the sharp rise in electronics R & D over

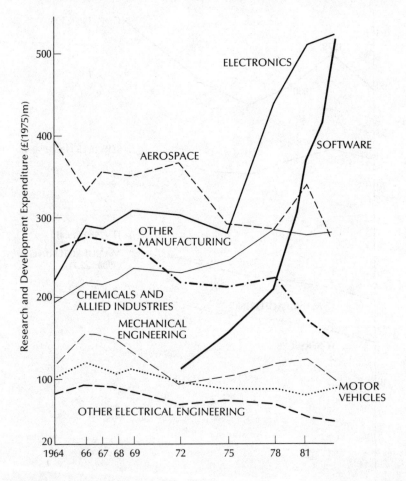

Figure 10.2 Intra-mural expenditure on research and development at 1975 prices
Source: CSO *Annual Statistical Abstract* as in Soete (1985) and table 10.7 for software

Table 10.7 Changes in research, development and software expenditure: UK, 1975 to 1983 (in 1975 prices and % of total)

	1975		1978		1981		1983	
	£ Million	%	£ Million	%	£ Million	%	£ Million	%
Manufacturing R & D	1293	86.8	1512	85.1	1555	76.6	1465	70.2
Chemicals	254	17.0	284	16.0	277	13.6	279	13.4
Mechanical engineering	103	6.9	118	6.6	111	5.5	99	4.7
Electronics	279	18.7	442	24.9	511	25.2	529	25.3
Other electrical	73	4.9	69	3.9	53	2.6	45	2.2
Motor vehicles	88	5.9	88	5.0	80	3.9	91	4.4
Aerospace	291	19.5	285	16.0	337	16.6	272	13.0
Other manufacturing	213	14.3	226	12.7	185	9.1	151	7.2
Non-manufacturing R & D	48	3.2	54	3.1	106	5.2	99	4.7
Software	149	10.0	210	11.8	369	18.2	524	25.1
Total R,D & S	1490	100.0	1776	100.0	2030	100.0	2088	100.0
IT-total (electronics and software)	428	28.7	652	36.7	880	43.3	1053	50.4

Source: Compiled from *British Business*, 18 January and 27 September 1985

Table 10.8 UK Government expenditure on R & D in electronics & information technology 1983–84

Sector	MOD	DTI	SERC	Other	Total £ millions
Office machinery	–	2.5	–	–	2.5
Electronic data processing equipment	47.0	25.6	0.1	–	72.7
Telegraphic & telephone apparatus		3.2	–	0.1	
Electrical instruments and control systems	336.0	17.2	0.1	0.7	365.5
Radio & electronic capital goods		6.7	–	1.3	
Passive electronic components		0.1	0.1	–	
Active electronic components	60.0	0.2	0.1	–	66.9
Electronic consumer goods		6.6	–	–	–
Total	443.0	62.1	0.4	2.1	507.6

Source: K. Guy (1986)

the last 20 years understates significantly the impact of IT because much of the technical change associated with computerization in industry and services takes the form of new software design and applications, which is generally not included in present definitions of R & D.

In figure 10.2 the trend in software expenditure (in 1975 prices) over the period 1972–83 has therefore also been drawn. By 1983, software expenditures were at about the same level as total electronics R & D expenditures, having grown even faster over the last 5 years. Taken together these sectors, corresponding relatively closely to the concept used earlier of the 'IT-producing' sector, amounted by 1983 to just over 50 per cent of what could be baptized as *R,D & S:* Research, Development and Software. As table 10.7 illustrates, in 1975 this figure was only 28.7 per cent. The information technology R,D & S total has increased over the period 1975–83 at an average annual growth rate of some 11.25 per cent. The contribution of software expenditure in this IT total is particularly worth noticing and points at a more general level to the increasing importance of services in bringing about technical change. Richard Barras (1983) has suggested that as some service industries, such as financial services, increase their capability in the design, development and maintenance of new software systems, they will become important sources of new innovative services used elsewhere in the economy, i.e. they will tend to become IT-producers, as well as users, in our terminology. His point is well taken and could well explain the dramatic growth in software expenditure, illustrated in figure 10.2.

With regard to the IT manufacturing producing side and in particular electronics, the concentration of total R & D on electronics may be a little higher in the United Kingdom than in most other countries. The emphasis on electronics R & D in the United Kingdom is primarily a reflection of the heavy government support for military R & D in electronics. As table 10.8 illustrates, nearly 95 per cent of total government support for electronics R & D originates from the Ministry of Defence. This 'military' R & D focus reflects itself also in the structure of the UK electronics and IT industry, which appears in its manufacturing 'IT-producing' part more geared towards the non-commercial, more sophisticated side of the industry unable to cope with the growth in civil domestic demand.

The importance of the military sector is actually brought out in the electronics output figures, where, as table 10.9 illustrates, by 1983 already more than 20 per cent of total sales could be accounted for by the Ministry of Defence's expenditure on defence equipment.

As we emphasized in chapter 8, one of the most specific features of IT, which elevates it, in contrast to many other R & D intensive sectors, to the rank of 'pervasive' technology is its widespread user impact. Complementing the data presented in chapter 8 with respect to the origin

Table 10.9 Defence equipment expenditure as a percentage of the sector's total sales (1979-84)

	1979–80	1980–81	1981–82	1982–83	1983–84
Ordnance	40.8	59.2	55.6	56.4	55.5
Mechanical engineering	1.6	2.1	2.0	2.3	2.8
Data processing	–	–	7.9	5.6	6.3
Electrical	–	–	2.4	2.2	1.9
Electronics	15.6	18.5	21.3	20.6	21.3
Motor vehicles	1.5	2.0	1.9	2.4	2.0
Shipbuilding	37.8	58.5	88.7	42.2	53.1
Aerospace	50.2	43.3	49.4	46.4	42.3
Instruments	7.6	9.2	11.8	8.4	7.4

Source: Calculated from *Business Monitor*

and use of significant innovations introduced in the UK over the period 1945–83, we present here some more recent innovation data developed in 1982–3 by the British Technology Group in collaboration with the *Financial Times* based on 'news' coverage, i.e. a sort of 'innovation' bibliometrics. The advantage of this data set is that it not only consists of a rather wide spectrum of reported 'innovations' (some 1500 per quarter) classified at a relatively detailed level of disaggregation but also it provides information as to the origin of the sector of the innovating firm and the sector of potential use.

The significant difference in diffusion user potential between the various sectors considered emerges quite clearly from the FT data. Sectors such as plastics, instruments and electronics see their innovations used over more than 30 sectors. Figure 10.3 represents for each sector considered a concentration/diffusion index based on a Herfindahl index but adapted to take into account the degree of sectoral diffusion.[1] Most 'diffused' appear the following: control systems (0.04), electronic capital goods (0.05), measuring instruments (0.08) and computer software (0.08). Most 'concentrated': construction equipment (∞), mining & extraction equipment (2.59), agricultural machinery (1.66) and textile machinery (1.11).

By and large, the results obtained in figure 10.3 confirm the user-pervasiveness argument with regard to IT. The importance in terms of diffusion of computer software rather than simply the manufacturing computer hardware emerges, however, also from the data presented here. As compared to the very low concentration index of computer software (0.08), the index obtained for computers is somewhat higher at 0.34 and for computer peripherals at 0.23. It seems to illustrate the overriding importance of user-interaction in the case of computer software, as compared to computer hardware.

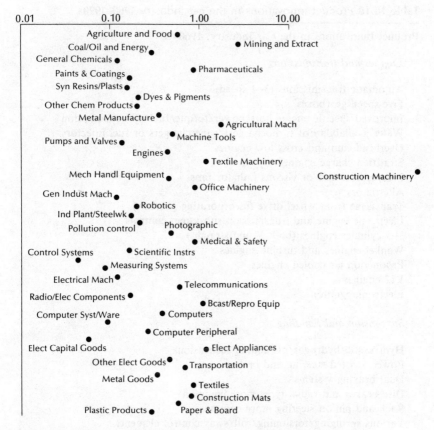

Figure 10.3 Technological user concentration/diffusion by sector

Figure 10.3 highlights also the importance of the IT control and measuring instrument sectors. These sectors have been, together with the electronics industry itself, among the first to witness their own 'electronification' (McLean and Rush, 1978; Freeman, 1982). These same sectors are now also at the heart of the further diffusion and use of IT equipment.

The point can be readily illustrated by looking in somewhat more detail at the underlying sectoral innovation data. In doing so we will rely heavily on the set of sectoral reports published by Gower. These reports point clearly to the significance of the shift towards IT innovations over the seventies. This change is illustrated in tables 10.10 and 10.11 for the automobile and transport industries and table 10.12 for textiles and clothing.

Whereas many innovations in the forties to sixties were based on the scaling up of mass and flow production techniques, taking advantage of cheap energy and petrochemical innovations, in the sixties to eighties the

Table 10.10 Product innovations in the car industry 1960–1990s

Product Innovations in the Car Industry, 1960–80

Engines and transmissions

Automatic transmissions (3–4 speed)
Five-speed gearboxes
Improved specific engine horsepower/torque/fuel/oil consumption
Wider availability of turbochargers/superchargers or fuel injectors
Overhead cam and crossflow engines
Stratified charge engines
Independent and/or viscous radiator fans
Alternators
Transverse front wheel drive (incorporating gearbox)
Electronic engine and transmission subsystem controls
3–5 cylinder engines (both V and straight)
Wankel engines and turbine engines
Production air cooled engines
V12 engines
Electronic ignition

Suspension and handling

Hydrolastic/hydrogas/pneumatic suspensions
Power assisted steering and brakes
Dual braking systems
Disc brakes and radial tyres more common
Rack and pinion steering more common
Various springing/torsioning/roll/sway control elements
More than one driven axle
Anti-lock braking system

Passenger and operator environment

Plastics and trim upholstery
Large distortion free safety glass areas/fewer blindspots
Heated rear windows and wipers
Window washers and variable cycle wipers
Halogen headlights
Sound deadening materials (mainly plastic)
Safety straps and padding
Collapsible steering wheels
Electronic instrumentation and controls
Printed circuit instrument panels
Glass fixing and bonding (direct bonding and flush glass)

(continued)

Table 10.10 *(continued)*

More Recent Innovations in Automobile Production Processes

Tooling:

Stamping

Larger presses (larger panels)
Pick and place machines between presses
Automatic loading and stacking at each end of the press line
Rapid die changeover times
Automated waste recovery

Body building

Multiwelders
Robotwelders
Pick and place machines and transfer lines
Automated monitoring of weld quality

Painting

Electro dipping and rust-proofing
Automatic paint sprayers
Rapid colour change
Robot paint sprayers

Powertrain

Automated machining
Computer controlled transfer lines
Large complicated engine and transmission Alcastings

Trim and assembly

Automated trim cutting
Portable power tools
Computer controlled transfer lines

Product Innovations in the Car Industry in the 1980s and 1990s

Powertrain

Electronic engine controls for performance and fuel efficiency
Electronic controls essential for future developments of IC and
    diesel engines
Electronically controlled cylinder cut-off devices
Electronic fuel management, injection, etc.

(continued)

Table 10.10 *(continued)*

Electronically controlled automatic or CVT transmissions
Integrated fuel, engine and transmission management to optimize the
  performance of each
Electronic water pumps, radiator fans, etc.

Suspension and handling

Electronically controlled anti-skid braking
Electronically controlled four-wheeled drive
Electronically controlled anti-slip acceleration
Electronically controlled anti-skid steering, including power steering and
  four-wheeled steering
Electronically controlled suspension, adjustable to load and conditions
Electronic suspensions
Electronically controlled collision prevention (sonar)

Passenger and operator equipment

Solid state electronic instruments
Rain sensitive electronically controlled wipers
Electronic fault diagnosis and system wear monitoring
Electronic maintenance scheduling
Trip scheduling, route guidance, navigation
Highway control devices (road tax)
Talking car (warnings, etc)
Electronically controlled restraint devices (airbags, etc)
Electronic climate control, electrically controlled air-conditioner
  compressors
Electronic entertainment devices

Note: This is an incomplete list to convey the pervasiveness of electronic related innovations
in the automobile in the 1980s.
Source: Compiled by Paul Gardiner, SPRU for chapter 2 (Vehicles) by D. T. Jones in (Ed.)
C. Freeman, Vol. 4, *Technological Trends and Employment.* Gower, 1985.

emphasis shifted to computerization of many processes and systems.
The effects of the new information technology on product design and
innovation can be seen both in the case of the car industry and in research-
intensive industries such as scientific instruments (table 10.13) which in
terms of its IT use has now become part of the IT sector.

By Way of Conclusion:
IT's Pervasive Productivity Growth Potential

We have tried to illustrate in this chapter that the new information
technology is likely to open up many new areas of 'trade', particularly

Table 10.11 Innovations in the UK transport sector 1940s to 1980s

1 *System innovations and economies of scale to take advantage of low cost oil-intensive technology (1940s to 1960s)*
Containerization
Unitization
Roll-on, roll-off
Oil and gas pipelines
Dieselization of railways

Scaling up size of trucks (to 38 tons)
Scaling up size of aircraft (from DC3 to 747)
Scaling up size of tankers (from 20,000 dwt to 500,000 dwt)
Scaling up of port facilities to handle large tankers, roll-on, roll-off, containerization and unitization

2 *Information and control innovations to take advantage of increasingly low cost electronics and computing (1960s to 1980s)*
Radar and computer-controlled airport traffic
Computerized airline booking systems
Aircraft instrument landing systems and flight control systems
Computerization of railway marshalling yards
Computerization of railway signalling systems
Computerization of road haulage and delivery systems
Unmanned trains
Tachograph
Computerization of travel agencies
Computerization of road traffic control systems

Source: Derived from working paper by I. M. Brodie, 'Transport', TEMPO Sector Study, SPRU (mimeo), March 1984

in the service sector. Over the last 5 years, employment in the United Kingdom in IT, including both an IT-producing and a predominantly IT-using sector, increased by some 163,000. On a similar definition basis, it was estimated that IT represented some 22 per cent of total GDP in 1984; a more important contribution to GDP than manufacturing. With regard to what was baptized research, development and software, IT, taking into account software expenditures, represents now more than 50 per cent of total UK R,D & S. Finally with respect to international trade, it was argued that IT will open up an increasing gamut of services to international trade, possibly lifting to some extent the manufacturing balance of payments constraint on recovery growth in the United Kingdom.

Overall, then, the emergence of IT over the last decade as a new pole of economic growth in the United Kingdom is clear cut and by any standard impressive. However, the potential impact of IT on the growth and 'efficiency' of the rest of the economy will, we would argue, be even more

Table 10.12 Innovations in the UK textile and clothing industries, 1940s to 1980s

1 *Radical innovations in the materials used in the industry, based on flow-production petro-chemical processes developed in the chemical industry (1940s to 1950s)*

Nylon
Terylene
Courtelle
Tricel
Polypropylene
Non-wovens

2 *New and improved yarns, fabrics etc. based on new synthetic materials and mixtures with traditional fibres (typical examples) 1950s to 1960s*

Crimplene
Propylene tape
Plastic net and net-like fabrics
Lurex
Nylon yarns for carpets
Tufted carpets

3 *Radical innovations in textile machinery increasing speed and reliability for basic textile processes (spinning, weaving, finishing) and taking advantage of availability of new synthetic materials (1950s to 1970s)*

Sulzer loom
Shuttleless loom
Water jet loom
Open end spinning
Repco self-twist spinning machine
Adhesive laminating machines
Crimping machines
Automatic doffing

4 *Combined effects of radical and incremental innovations in basic textile processes (1940s to 1970s)*

Spinning (lbs per spindle hour) from 0.019 in 1940 to 0.065 in 1970.
Extrusion of polymers from 7 kg per hour in 1950s to 70 kg per hour in 1970.
Drawing speeds from 80 to 350 m per minute. Maximum spindle speeds in texturing machines from 20,000 rpm to 800,000 rpm.
Warp knitting machine speeds from 200 courses per minute in 1950 to 2000 courses per minute in 1978. Weaving speeds increased by over 50 per cent in 1960s.

(continued)

Table 10.12 *(continued)*

5 *Instrument innovations (including electronic instruments) contributing
    to quality control, reliability control and environmental control (1970s)*

Shower testers
Fibre strength testers
Stitch damage tester
Colour difference meter
Moisture analyser
Fabric inspection machines
Weigh stoppers

6 *Computerization (1960s to 1980s)*

Computerization of stock control
Computerization of looms
'Computaknit' and other computerized knitting systems
Continuous computerized finishing integrating dying and
    finishing techniques
Electronic pattern-making

7 *Major technical innovations in clothing industry (1970s to 1980s)*

Automatic contour seamers, profile stitching machines and numerically
    controlled sewing machines

Laser cutting (guided by computers and cutting fabric at high speeds with
    high accuracy and reducing material losses)

Numerically controlled cutting devices

Ultrasonic sewing – ultrasonics used to create frictional bonding between
    layers of thermoplastic cloth. No thread and no adhesive required.

Computerization of sales analysis and forecasting, process inventory and
    work flow management, linked to numerical control equipment.

Sources: Based on data from:
L. Soete, 'Textiles', chapter 3 in (Ed.) K. Guy, *Technology Trends and Employment*,
Volume 1, *Basic Consumer Goods*, Gower Publishing Co., London, 1984
H. Rush and L. Soete, 'Clothing' chapter 4 in (Ed.) K. Guy (as above); and *'Technical Change
and its labour impact in five industries'*, Bureau of Labour Statistics, Bulletin no 1961,
Washington DC, 1977.

important and emerge as an even more striking feature of the pervasive
impact of the new technology.

The possibility to collect, generate, analyse and diffuse enormous
quantities of information at minimal cost affects the productivity of all
factor inputs. It is this feature which distinguishes IT so clearly from

Table 10.13 Major technological changes in the scientific instrument industry

Technology	Description	Diffusion
Microprocessors and computers	Microprocessors and computers are used either in providing the processing power of single instruments or systems generally as a basis of management information systems.	Extremely fast since the mid 1970s. In the mid 1980s hardly any instrument will fail to incorporate some form of micro-electronics.
Computer aided design	The productivity of designers in the drawing office in the design of PCBs[a] increases by a factor of 3.	Diffusion was relatively slow until the late 1970s. Since then the rate has increased substantially and although cost is a substantial barrier CAD is likely to be an essential design tool in every office of the late 1980s.
Automated insertion and test equipment	The productivity of component assembly of PCBs[a] increases dramatically. ATE[b] insures marginal rejection rates assemblies.	Automated insertion is increasingly standard in volume PCB assemblies. ATE is already an essential aid for a PCB assembler.
Lasers	A basic CO_2 laser consists of a gas mixture in a discharge tube as the lasing medium, a high voltage power supply and an optical cavity of two mirrors.	Early stages of diffusion. Vast potential in medical and analytical application.
Fibre-optics	Signal transmission system with a potential for substituting the copper wire.	Early stages of diffusion but some fast growth applications such as medical enteroscopy and process control cabling.

Source: Rendeiro (1985)

[a] PCB, Printed Circuit Board
[b] ATE, Automated Test Equipment

automation. As was highlighted in chapter 4, some of the most significant productivity gains linked to the introduction of IT relate to more efficient inventory control, as well as significant energy, materials and capital savings.

Whereas much qualitative evidence exists about some of these savings, relatively little empirical evidence exists with regard to IT's non-labour productivity gains. As long as no systematic data exist about IT-investment outlays, it is however very difficult to get even an impression from the available sectoral productivity data of some of these productivity effects.

Figure 10.4 presents some rare US data with regard to the sectoral allocation of IT sales. What emerges from figure 10.4 is that the main users of IT equipment are, as already hinted at above, in the first instance the sectors which are at the actual origin of IT equipment, in particular the computer, communications (including electronics) and instruments sectors. In other words, if anything, the 'pervasive' user productivity growth potential of IT should, at least for the past period, emerge in the first instance from the productivity data of the IT producer sectors themselves.

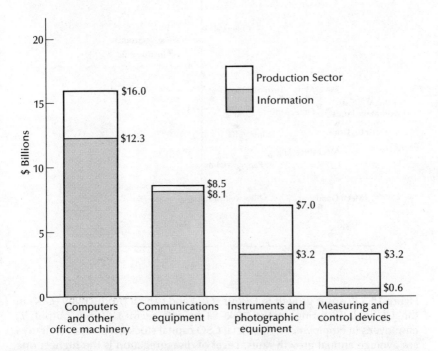

Figure 10.4 Who uses high-tech equipment? (allocation of private domestic final shipments in 1982 in the USA).
Source: Roach, 1986

Based on the data presented in chapter 8, limited though to manufacturing and covering the full period 1948–84, figure 10.5 brings together the postwar growth rates in labour and capital productivity in the 18 industrial sectors for which 'official' capital stock estimates were available. Growth in labour productivity is represented on the vertical axis and growth in capital productivity on the horizontal axis.

As already hinted at in chapter 8, figure 10.5 suggests that only the 'core' IT-producing sectors witnessed a simultaneous growth in both labour and

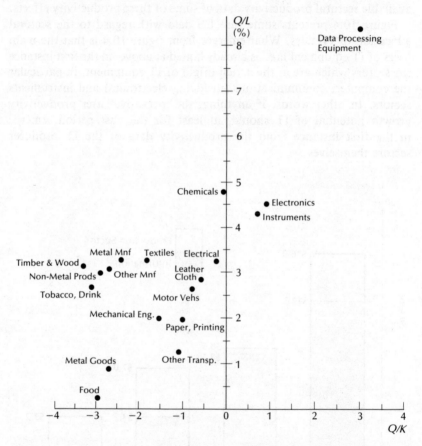

Figure 10.5 Postwar growth in labour (*Q/L*) and capital productivity (*Q/K*) in the UK manufacturing sector (1948–84 by SIC sector.) *Q*, net output; *L*, employees in employment; *K*, official CSO capital stock estimates. All figures are average annual growth rates. Level of disaggregation is the highest one available in terms of capital stock estimates, except for electronics, which has been estimated on the basis of Soete and Dosi (1983). All sectors have been normalized to their 1980 SIC definition (for more detail see chapter 8)

capital productivity. It lends support to the argument that IT is indeed pervasive in its productivity impact, at least with regard to the traditional factor inputs labour and capital. As figure 10.5 illustrates, the debate about the employment impact of information technology centres on two propositions which are quite distinct from the traditional arguments about 'technology-induced' employment displacement in most manufacturing sectors. First, information technology is not so much 'labour saving' but leads to increases in both labour and capital productivity, and could thus well be considered as 'capital-saving' in the Harrod sense. Second, as a consequence it will lead on the one hand to a more rapid substitution of capital for labour, affecting in particular those sectors that have witnessed little labour displacement so far (e.g. services) and relatively labour-intensive sectors (e.g. clothing); while, on the other hand, it will lead to substitution of IT equipment which we would like to refer to as 'intelligent capital' for older, electromechanical capital (including some first or even second generation electronic capital equipment).

This will be particularly the case in those highly capital-intensive sectors where investment costs are one of the major growth inhibiting factors, and where the economies of scale linked to large-scale investment are a major factor in preventing firms from acquiring the required flexibility in periods of low output growth and uncertainty, leading to either large-scale under-utilization of capital or significant 'scrapping' of productive capital.

Underlying the growth in capital productivity in some of the most characteristic IT sectors, such as electronic computers, electronics and instruments one observes, we would argue, the rapid diffusion of semi-conductor technology. It should be remembered that in chapter 8 it was found that this sector (MLH 364) was the main originating sector or 'innovation supplier', in these three IT sectors (see table 8.2). Its impact, we would argue, can be compared to a process of more or less continuous material-saving technical change.

This process has in many ways been responsible for the specific capital-saving nature of technical change in the IT industries themselves and in particular in the computer, electronics and instruments industries as the most important users of semiconductor output. The extent of the capital-saving bias or more precisely of the simultaneous growth in labour and capital productivity provides in our view further support for the argument that the further diffusion of such 'intelligent capital', computers in particular, might well have a significant overall growth impact on the economy.

Note

1 For more detail see Soete (1986).

For each origin sector, the index was calculated as $\dfrac{\Sigma\ (s_i)^2}{\log n}$,
where s_i is the sectoral user share and n the number of user sectors. The higher the index the more concentrated or localized the user impact. In the limit with no user impact outside one's own sector

$$s_i = 1 \text{ and } n = 1 \text{ or } \frac{\Sigma\ (s_i)^2}{\log 1} = \infty.$$

This is e.g. the case for construction equipment. The lower the index, the more widespread the diffusion user impact. In the limit, diffused equally over all 43 sectors,

$$s_i = \frac{1}{43} \text{ and } n = 43, \ \frac{\Sigma(\frac{1}{43})^2}{\log 43.} = 0.0003$$

IV

POLICY IMPLICATIONS

Having presented in the previous chapters some of the scattered empirical evidence with regard to employment growth and decline in the various sectors of the UK economy, we turn now in this last part to a more qualitative discussion of the likely trends in future (un)employment growth in the United Kingdom and a discussion, albeit brief, of some of the policy implications of our analysis. In part II we presented, as part of the methodological discussion, some employment forecasts for the UK manufacturing sector. These pointed to the limited employment growth potential in manufacturing and the existence of a severe 'capital-shortage' constraint on future employment growth given present trends in technical change. The analysis was limited to manufacturing mainly for methodological reasons. With respect to many services, and even more so to the newly emerging IT sector, a capital vintage simulation exercise, apart from the lack of much of the required data, would by and large be inappropriate.

Although it presents various quantitative projections the analysis in chapter 11 starts therefore from a more qualitative discussion of the likely future aggregate employment trends in UK employment. This analysis, based on the empirical evidence presented in chapters 9 and 10, points to the significant employment growth potential in many service sectors, including those related to IT. However, at the aggregate UK level, it appears that this growth potential is unlikely to be sufficient to bring about a significant decline in present levels of unemployment without policy changes. The need for a reduction of 'mass' unemployment emerged however, from the outset (chapter 1), as one of the most essential requirements for social peace and economic stability, and, as the title of our book suggests, the attainment of some notion of 'full' employment is one of the major aims we set out to illustrate in our study. Where and how employment growth could be brought about, sufficient to achieve

221

such reduction to 'acceptable' levels in unemployment, is therefore discussed in chapter 11. This highlights the crucial importance of the policy implications of our analysis discussed in chapter 12.

Clearly, as both the theoretical and empirical analyses have brought to the forefront, and as so many contributions in the field of labour economics have illustrated, there are no easy, immediate policy solutions to the present levels of 'mass' unemployment. Our analysis, starting from the importance of technical change in bringing about employment growth and displacement, points to a somewhat different set of policy conclusions from those generally found in traditional labour economics. First, with regard to wage and income policies, our analysis points to the importance of wage flexibility, rather than the mere observation that present levels of wage settlement are above the 'full employment' equilibrium wage. As already brought out in the so-called efficiency wage literature, there is nothing illogical in firms paying out higher wages once training and exit costs are taken into account. Secondly, we emphasize to a far greater extent the crucial importance for future employment growth of (re)training policies particularly with regard to IT-related skills. In contrast to many contributions in labour economics using evidence based on the trend in the aggregate, unemployment–vacancies relationship, our analysis points to critical skill shortages. Third, the concept of change in techno-economic paradigm which we would associate with the newly emerging information technology brings to the forefront more than in most other analyses, the essential need for institutional change, covering a wide array of economic, industrial and innovation policies.

11

Quantitative Analysis of the Future of UK Employment

John Clark and Christopher Freeman

Many forecasts of future levels of employment and unemployment in the United Kingdom are produced on a regular basis by well-established research groups whose work is centred around a macroeconomic model of the economy. In this chapter we shall look at some of these forecasts and make some estimates of our own. The major objective is not, however, to present yet another set of numbers; we believe that work on fully elaborated econometric models of the United Kingdom is already well subscribed and that the most useful contribution we can make is to focus on particular issues related to the employment implications of changes in technologies incorporated into the system of production. The approach is necessarily less all-embracing than that typically used to obtain projections of employment, but is designed to illustrate important problems and policy issues which need to be resolved if the apparently intractable problem of high unemployment is to be overcome.

Table 11.1 shows a set of forecasts of unemployment in the United Kingdom by various research teams in 1983–5. It is notable that the expectation of a significant decline in unemployment towards the end of the eighties was the exception rather than the rule, and that the basic economic philosophies of the 'optimistic' groups are clearly distinguishable from those of the more 'pessimistic' teams; it is perhaps hardly surprising that a priori opinions and beliefs about how the economy works condition the content and hence the prognoses of economic models. Despite the increasing sophistication and empirical rigour of forecasting models, the subjective element remains high. An example of this is the inclusion of equations relating to factor substitution in the 1985 version of the Treasury model, following similar assumptions incorporated in the LBS and

223

Table 11.1 Projections of UK unemployment (millions)

	1984	1985	1986	1987	1988	1989
National Institute of Economic and Social Research[a]	3.1	3.3	3.4	3.5	3.5	
——— [b]		3.2	3.3	3.4	3.3	3.3
Institute for Employment Research, University of Warwick[c]	3.2	3.1	3.0	3.0	3.0	
London Business School[d]	3.0	3.2	3.1	2.9	2.8	
Liverpool Research Group in Macroeconomics[e]	2.8	2.4	2.1	1.8	–	
——— [f]		3.2	3.0	2.8	2.7	2.4
City University Business School[g]	3.0	2.9	Average 1.7 to 1990			
——— [h]	3.0	3.1	2.8	2.6	2.5	2.1

[a] *National Institute Economic Review*, November 1983; August 1984. Figures refer to December each year.
[b] *National Institute Economic Review*, May 1985, November 1984, Fourth Quarter.
[c] Institute for Employment Research, University of Warwick, *Review of Economy and Employment*, Summer 1983.
[d] LBS Centre for Economic Forecasting, *Economic Outlook*, Vol. 8, no. 9, Gower, June 1984.
[e] *Quarterly Economic Bulletin*, Liverpool Research Group in Macroeconomics, May 1984.
[f] *Quarterly Economic Bulletin*, June 1985.
[g] Economic Review, City University Business School, Spring 1984. 'Medium Oil' projection. Unemployment 1991–95, 2.0 million; 1996–2000, 1.8 million.
[h] *Economic Review*, Spring 1985. Figures inputed from quoted percentage unemployment rates, using Department of Employment labour force statistics. Unemployment projected to fall to under half the 1985 level during the 1990s.

Liverpool models. Even so, it seems that none of the teams expected a reduction to even the highest unemployment levels observed in the 35 years following the Second World War in the foreseeable future.

The model underlying the National Institute forecast is based on Keynesian principles. NIESR related their earlier projection of unemployment rising from 3 to 3.5 million towards the end of the eighties (*Economic Review*, November 1983, p. 48) directly to an expectation of continuing deficiency of aggregate demand rather than to 'structural' factors such as a lack of flexibility in the labour force (table 11.1). They did, however, acknowledge the possibility of supply-side problems arising from continuing high unemployment: 'The ultimate consequence could be a lower level of supply potential as the capital stock adjusts down to a lower level of expected output. . . .' (NIESR, 1983, p. 48)

The November 1984 *Economic Review*, while suggesting a slightly less pessimistic outlook for employment (table 11.1) highlighted further the question of supply capacity, citing CBI data showing that capacity utilization in 1984 had risen to near the 1979 level despite the dramatic rise in unemployment since then.

The model used by the Warwick group is a modified version of that developed by the Cambridge Growth Project team (see chapter 5; the latter group is now privatized as Cambridge Econometrics). This model also has an essentially Keynesian structure, with a focus on the 'real' sector which reflects not so much a belief that financial and monetary flows are unimportant but more the stage reached in a research strategy orientated towards assessing medium-term employment prospects and identifying their structural characteristics. Also, 'to make these [exchange rate and interest rate] variables endogenous[1] in the medium-term context requires further work on their determinants in order to produce credible explanatory relationships' (Lindley, 1980). This model is unusual in the degree of disaggregation employed; the economy is divided into some 40 sectors, compared with the three or four used in most other macroeconomic models. As we have already seen a central feature of the model is an input–output framework incorporating flows between the various sectors.

The London Business School model laid greater stress on monetary variables and the existence of equilibrium long-run relationships (Ball, Burns and Warburton, 1979). This model has been described as 'mildly monetarist' (Holden, et al., 1983, p. 106) in contrast to the more explicitly monetarist position adopted by the Liverpool Research Group in Macroeconomics (LRGM). The latter group is unequivocal in laying the blame for high unemployment on the upward pressure in real wages. In their view, 'this Government rightly is committed to reducing unemployment by removing obstacles to the proper functioning of the labour market' (Minford, 1984), which included, in their opinion, immediate abolition of Wage Councils and repeal or weakening of the Employment Protection Act.

In an analysis of unemployment between the Wars, Matthews, a member of the Liverpool team, concluded that unemployment benefits played a large part in preventing downward wage adjustments and hence (he argued) in keeping unemployment at a high level. Although a reduction in benefits alone would not have been sufficient to reduce unemployment, 'the lessons for the 1980s are however there to be learned' (Matthews, 1984). By implication, current unemployment benefits are above the level required to give unemployed individuals the incentive to seek employment at wage rates which potential employers are able to offer (see chapter 1). As Layard and Nickell (1985) have pointed out, the very high employment gains derived from reducing benefits result from the fact that benefits and union power are the only trended variables in Minford's wage equation – given the assumptions used the diagnosis is hardly surprising. In the medium term, the LRGM group believed that such 'imperfections' in the operation of the market would be overcome and unemployment would consequently fall. More recently though, their forecasts were revised dramatically upwards from some 1.8 million forecast for 1987, to 2.8 million (see table 11.1).

The City University Business School also suggested that unemployment would fall significantly in the late eighties owing to low growth in real wages. Their model incorporated the concept of a 'natural rate' of unemployment (which they estimated at around 7 per cent, corresponding to a level of around 1.75 million in the late eighties). To achieve this level, a growth in real wages of well under 1 per cent per annum up to 1990 was considered necessary (CUBS, *Economic Review*, Spring 1984). As with Liverpool, the City group appeared reasonably confident that their preconditions for a reduction in unemployment would be achieved.

In general, the 'Keynesian' models were more pessimistic than those with a chiefly 'supply-side' orientation (although, as table 11.1 shows, the divergences had become smaller by 1985). If employment is determined mainly by the level of aggregate demand, which in turn is particularly sensitive to the volatile nature of investment demand, prognoses of depressed levels of investment expenditures necessarily imply high unemployment. If, on the other hand, disequilibria in the labour market represent the major cause of high unemployment, it is not unreasonable to expect that, sooner or later, market imperfections will 'work themselves out' leading to a progressive reduction in unemployment. The more flexibility that is assumed – such as in the allowance for factor substitution in the 1985 Treasury model – the greater the potential or likelihood of labour market clearing.

A focus for our own analysis has been in the level and nature of spending on fixed capital. It is important in that it increases both aggregate demand and aggregate supply, and hence can help relieve more than one constraint to full employment. It is of particular interest to us in that technological change has an important influence both on the magnitude of investment (to the extent that innovations provide new opportunities for profitable exploitation) and in that newly installed fixed equipment has employment-generating potential (to an extent dependent on the nature of the technologies embodied in it). It is essential to remember that new investment has this broadening aspect as well as a deepening aspect.

The extent to which technical change acts as a stimulus to investment demand is difficult to assess statistically, partly because of the difficulties with measures of technical progress (chapter 5). However, the theoretical considerations in chapters 2–4 and much of the more qualitative examination of the impact of technical change on employment that we have undertaken in chapter 10 and in our sector studies strongly supports the view that technical change has an important influence on capital spending.

One illustration of this is the increase in capital intensity in many service sectors, and especially in the financial services associated with the introduction of information technology (see chapter 9 and volume 5 of the Gower series). Many other examples are shown in the manufacturing sector studies (chapter 8 and volumes 1–4 of the Gower Series), such as the growth of the synthetic materials industry in the fifties and sixties (volume 2), the growth of the electronics industry over the whole postwar period (volume 3), and the introduction of CAD, CNC and related technical changes in the engineering industries (volume 4).

From these and other sector studies it is clear that new technology affects investment and employment in two ways. First, it generates changes in the scale and direction of new investment by offering new opportunities for profitable exploitation of new and improved products and processes and in response to competitive pressures. Secondly, the newly installed 'vintages' of equipment offer employment opportunities, which vary with the embodied technology.

However, the necessary stimulus to new investment and to this more rapid diffusion of new technology would not occur without major changes in policy. Without such changes our estimates also suggest little or no improvement in the level of unemployment, since the probable increases in private service employment would probably be offset by further decline in other sectors and by the expected growth in the labour force.

We first of all examine the likely trends in unemployment up to 1988 on the assumption of no major changes in policy and then discuss the prospects from 1988 to 1995 on the assumption of policy changes designed to promote an investment boom based on the rapid diffusion of new technology.

The Employment Future with Unchanged Policies

The outlook for manufacturing employment is for little change. Evidence presented earlier in this book suggests that the amount of unused capacity in the economy has declined in recent years despite high and increasing levels of unemployment. It is suggested that a sustained period of low demand and high unemployment, as in the seventies, might be expected to be accompanied by a reduction in the level of productive capacity, after which any upturn in demand would fail to reduce unemployment owing to a supply-side constraint. As shown in chapter 7, the capital–vintage model described in chapter 6 can be used to examine the conditions under which such a problem could arise.

The overall conclusion from that chapter was that a capital constraint to growth could well occur in the United Kingdom in the medium term (late eighties, and early nineties), but this could be overcome by a combination of higher capital productivity associated with new technology (as discussed in chapter 8) and of higher levels of investment in manufacturing and services. This is one reason that we believe that policy changes designed to stimulate investment are required. However, the results already presented suggest that only a very modest rise in employment in manufacturing can be expected even under these favourable assumptions.

The overall conclusion from the analysis of various alternative future scenarios (chapters 7 and 8) for UK manufacturing employment in the next 5–10 years is that it is very unlikely that it will return to the level of the sixties. Even to maintain present (1986) levels of manufacturing employment or achieve a small increase would require high levels of investment and a high rate of technical change to maintain international competitiveness and improve capital productivity. We therefore assume no change in manufacturing employment up to 1988 and little prospect of any substantial increase up to 1995. This situation might of course change in the event of large-scale adoption of protectionist policies, as in the thirties, or an extraordinarily strong boom in capital goods, or more intensive re-armament.

The methodology used in our model is not, we believe, appropriate for an analysis of technology and employment in the non-manufacturing (principally service) sectors of the economy, although it is relevant to the 'network services'. A more qualitative approach along the lines of chapters 9 and 10 is more practical for the other sectors, where relationships between labour and capital are difficult to specify and where useful measures of output are often hard to define. The discussion here is limited to a brief assessment of the possible future employment generation in the service sectors on the assumption of unchanged policies up to 1988. The feasibility of large-scale employment generation in services thereafter is assessed in the light of the conclusion of the earlier chapters and other evidence.

Table 11.2 shows recent (July 1985) official projections of the civilian labour force in Great Britain to 1991. The average growth rate of the seventies (140,000, or 0.5 per cent per annum) was followed by a decline 1980–3, with growth of half a million in the following year; the labour force is expected to increase 1985–9, but will then level off. The main reason for the projected increase to the end of the eighties is the expectation of a rise of nearly half a million in the population of working age; much less predictable are the levels of activity rates which are also required for labour force estimates. In general it is assumed that high unemployment reduces activity rates (with, for example, less incentive for some groups of individuals to seek work, and a greater propensity for early retirement). The Department of Employment labour force projections in table 11.2 are based on the assumption that unemployment (the number of people claiming benefit) will remain at about 3 million throughout the period.

Table 11.2

	1980	1981	1982	1983	1984	1985	1986	1987	1988	1989	1990	1991
Labour Force (millions)[a]	26.2	26.2	26.0	25.9	26.4	26.6	26.8	27.0	27.1	27.2	27.2	27.2

[a] *Department of Employment Gazette*, July 1985. Civilian Labour Force, Great Britain.

The projected increase in the labour force up to 1988 suggests that employment in service and other non-manufacturing occupations could need to grow by about 300,000 if unemployment is not to increase over the period from 1986 to 1988. Where could new employment be generated to prevent any further increase in unemployment and to reduce the huge total below its present level?

It is easier initially to identify sectors where employment gains are extremely unlikely to occur. In this category we would include agriculture and three of the nine service industries identified in chapter 9 – utilities, transport and public administration.

The agricultural industries account for under 2 per cent of the working population, a proportion which has been declining steadily. Dramatic reductions in employment, however, seem as unlikely as significant gains – a continued gradual reduction in agricultural employment is almost inevitable. Because of its relative predictability and small size, agriculture has received very little attention in this volume. The same applies to mining, where, on the basis of present policies, some further decline also seems inevitable.

The network services discussed in chapter 9 (utilities, transport and communications) are, as pointed out there, akin to manufacturing in being highly capital intensive, and have shown, in aggregate, a similar pattern of employment loss as manufacturing. For utilities and transport, there are no indications of an upsurge of demand required to reverse this trend. For communications, it is possible to imagine large-scale investment projects in telecommunications which could restore employment growth in that subsector, so this is excluded from the list of sectors facing almost certain employment decline.

Regarding public administration, reductions in employment seem certain given current Governmental policies. There are plans for a gradual reduction in the number of civil servants to around 600,000 by 1988, and the targets should be comfortably met – they are far less ambitious than the 14 per cent decline achieved between the 1979 General Election and 1984. Pressure on local authorities to reduce manpower will continue. Job losses from the abolition of the GLC and other metropolitan councils were also inevitable. Other areas of public administration – justice, police and fire services – are relatively small and unlikely to see major changes in manpower levels.

A second group of three subsectors – construction, education and health – are also still vulnerable to significant job losses for the remainder of the eighties, but the situation is far from clear-cut in these sectors. The fortunes of construction are strongly dependent on the overall level of activity in the economy and, in particular, would be expected to fare better with relatively strong growth in manufacturing industry. The education and health sectors would seem to be rather better placed in terms of employment potential – these are areas where demand for the services concerned is far from satiated, and where in an ideal world almost indefinitely large resources could be usefully committed. They are also the sectors with the largest growth rates in employment over the postwar period. Against these factors, it is noteworthy that employment in education in 1985 was at about the same level as in 1976, while that in health has continued to grow over the last six years, but at only a fraction of the rapid growth rates achieved in the seventies. Being primarily state controlled, these labour-intensive sectors seem unlikely to undergo rapid employment expansion under a government committed most of the time to reducing public expenditure.

This leaves private services as the most probable source of any closing of the gap between labour supply and demand, but before assessing their potential for this we can estimate the size of the 'gap' by assuming that the historical decline in employment in agriculture, utilities and transport continues in the future; that the decline in public administration also continues according to Government policy: and that a constant level of employment exists in the second group of service subsectors (communications, construction and education and health).

These assumptions imply a further loss of over 100,000 jobs by 1988 in the primary industries and services combined (excluding private services). When put together with the earlier projections of manufacturing employment, the tentative conclusion is that, under present policies, about half a million new jobs would need to be created in private services between 1985 and 1988 if the official measure of unemployment is to remain around present levels. More recently (in 1986), there are some signs of relaxation of these policies.

If the average growth rate of private services employment over the period 1973–83 was sustained from 1985 to 1988, employment would increase by some 0.4 million in these sectors. We would expect this growth to continue but would conclude that, under present (1985) policies and industrial structures, the United Kingdom will be fortunate if unemployment does not increase over the remainder of the decade, taking into account the projected increase in the labour force.

We believe, however, that this outcome is by no means inevitable, and that with appropriate policy changes service employment can show very substantial growth over the coming decade. These policy changes are the subject of the final chapters, but we first indicate in the concluding section of this chapter those areas where there could be scope for further substantial increases in employment by 1995.

The Scope for Generating New Employment with Policy Changes

The Department of Employment's rule of thumb (*Employment Gazette*, February 1984), is that a reduction of 100,000 in unemployment would lead to a rise of 35,000–40,000 in the labour force. It is for this reason that the number of extra non-manufacturing jobs that need to be created to reduce unemployment below 1 million by 1995 would be well over 3 million and perhaps closer to 4 million.

Any speculation about non-manufacturing sectors in which such a growth in employment opportunities might arise is necessarily hazardous, but some suggestions can be made on the basis of the analysis in chapters 9 and 10.

It is important to realize, first of all, that huge though this task appears there are historical precedents in the United Kingdom for a high rate of job creation. Between June 1971 and June 1979, for example, even though economic growth was certainly not spectacular, employment in the service industries (1980 SIC classes 6–9) increased by nearly 2 million (table 11.3).

Table 11.3

		Employment in service (SIC 6–9) Thousand		
	1971	1974	1979	1985 [June]
Wholesale Distribution and Repairs 61/62/63/67	964	1023	1102	1337
Retail Distribution 64/65	1951	2048	2133	2153
Hotels and Catering 66	691	808	938	1041
Transport 7 exc. 79	1092	1031	1039	847
Postal Services and Telecommunications 79	435	434	413	419
Banking, Finance, Insurance 8	1318	1473	1638	1932
Public Administration 91 and 92	1733	1865	1947	1813
Education 93 and 94	1260	1450	1591	1669
Medical, Veterinary Services 95	939	1028	1186	1317
Other Services	979	1057	1251	1238
TOTAL:	11361	12217	13239	13439

1980 SIC – Figures relate to June each year
Source: *Department of Employment Gazette 96/7/8*

During the recovery from the deep depression of the thirties, employment in the United Kingdom grew by 2.4 million between 1932 and 1937. About half of the growth at that time was in manufacturing employment and for reasons which have been explained, this is unlikely to be repeated. However, taking into account the structural changes in the UK economy over the past 50 years, there is no reason in principle why it should not be possible to create 3 million jobs in 7 years in the service and construction industries (table 11.4). It is also certainly not inconceivable that an investment and construction boom could actually be strong enough to generate some increase in manufacturing employment itself, if the UK electronics industry can solve some of its structural problems.

The main areas of employment where it seems quite plausible to project substantial growth in the late eighties and early nineties are the following:

(1) The 'IT sector' described in chapter 10, and especially those IT producer and user services concerned with the diffusion and widespread application of information technology, i.e. software houses, consultancy groups, professional and design services, contract research, financial and marketing services, new types of publication and printing services. In this area skill shortages are likely to be the main constraint rather than capital shortage. For this reason the parallel expansion of education and training is essential.

(2) Income-elastic private services, which although also affected by capital-embodied technical change, nevertheless are, in very many cases, based frequently on a personal care or craft element, and are therefore likely to remain very labour intensive. These include catering, tourism, the boutique element in distribution, repair services, handicrafts, private tuition and training, and private health care. These have shown considerable employment growth, which is likely to continue within an overall expansion of the economy (see Gower Series, volume 5). In most OECD economies they show a strong upward growth.

(3) Public social services. Growth or decline in this area is heavily dependent on public policy and expenditure on education, health and social services. There is a very strong case for substantial expansion of education and training, based on the UK's relative weakness compared with other OECD countries, and the continuing problem of skill shortages and adaptability of the labour force. Education and training may legitimately be considered as a form of long-term infrastructural investment, as well as a public social service. The effects on unemployment are felt both through employment of teachers and tutors and through the people being taught.

(4) Construction. This sector too depends a great deal on public policy. Both the CBI and the TUC have advocated a major programme of infrastructural investment and there is also a case for a greatly increased public housing and urban renewal programme. The traditional Keynesian approach should be permeated by the need to provide infrastructure for information technology and to use IT in many other types of investment, e.g. combined heat and power schemes, domestic energy systems. There is scope for a major expansion of construction employment, as in the thirties and fifties.

The combined employment generation effect could amount to between 2.6 and 3.6 million new jobs and places in education, giving a 'central' estimate of 3.1 million (table 11.4).

This estimate of just over 3 million jobs and education places which might be created by 1995 is by no means unrealistic. As we have seen, nearly 2 million jobs were created in services between 1971 and 1979 when construction employment was declining and the rate of economic growth was relatively slow. Nor is it unrealistic in terms of international experience.

However, it is heavily dependent on assumptions about public policy in relation to construction and publicly provided social services, education

Table 11.4 Potential growth of employment, education and training places in UK economy from 1988 to 1995 on the assumptions of major changes in public policy and favourable growth in the world economy

	(millions)
IT-Producing and -using services (including financial, business, computer, professional, engineering)	0.5 to 0.7
Personal market services and other private services (including distribution)	1.0 to 1.3
Social services and administration (including private education and health)	0.3 to 0.4
Construction	0.4 to 0.6
Total new employment	2.2 to 3.0
Numbers of people taken into education and training and community projects	0.4 to 0.6
Total	2.6 to 3.6

and training. These are discussed further in chapter 12. Moreover, it would still probably leave some shortfall of jobs which it would ideally be desirable to create to reduce the level of unemployment to the Beveridge definition or to the lower levels achieved in the fifties and sixties.

One major advantage of the pattern of employment generation which has been suggested is that in no case should it involve a major increase in the strain on the balance of payments. Most of the areas suggested for expansion have very low import content and some could have positive balance of payments effects, especially producer services and IT-user services.

However, even taking this into account, a favourable trend in the world economy is an essential condition for this realization of UK expansion in employment and this can by no means be taken for granted. Clearly the United Kingdom has a very strong interest, in common with its European partners, in long-term solutions to the indebtedness problems of the Third World countries, and in ambitious re-structuring of world financial institutions to promote stable growth in the world economy.

But given a favourable trend in the world economy we would maintain that the type of increase suggested is certainly not unattainable. For the

Table 11.5 A US estimate of numbers of new jobs by 1995 (based on BLS estimates)

	No of jobs	% All new jobs
Retail trade	3,089,000	12.2
Business services	2,440,000	9.7
New construction	1,976,000	7.8
Eating and drinking places	1,583,000	6.3
Hospitals	1,461,000	5.8
Wholesale trade	1,149,000	4.6
Other medical services	1,024,000	4.1
Professional services	857,000	3.4
Education services (private)	514,000	2.0
Doctors' and dentists' services	502,000	2.0

Source: Perswick (1984) pages 30–32 (Valerie Perswick, The job outlook through 1995, *Monthly Labour Review*, November, 1984.)

Quoted in Etzioni, A and Jargowsky, P in *High Tech, Basic Industry and the Future of the American Economy.*

Human Resource Management, Autumn 1984, Volume 23, No. 3, pp. 229–40.

sake of comparison a forecast for the US economy of new jobs which
might be created by 1995 is shown in table 11.5. The conditions in the
US economy are of course very different from those in Britain, but the
point is nevertheless valid that a booming economy is quite capable of
sustaining a high rate of expansion in employment, both in private and
in public services, as well as in construction. So far from reducing this
possibility, new technology can greatly increase this potential, through
the type of multiplier effect described in chapter 10.

12
Policy Conclusions

Christopher Freeman and Luc Soete

It was not the purpose of this book to come up with a long and detailed list of readily implementable policy recommendations. Nevertheless, as the analysis in chapter 11 has illustrated, it is in our view essential to point to the wide range of organizational and institutional changes upon which future employment growth and a renewed upsurge of economic growth now depend. These changes are needed in all industrial countries but we have tried to indicate their specific relevance for the United Kingdom.

With regard to the impact of technical change on employment, there are no easy, immediate policy solutions. At first sight, there appears, as already indicated in the introduction and in the previous chapter, to be quite some scope in the United Kingdom for straightforward demand-reflationary policies at the macroeconomic level. On this we go along with the analysis of Buiter (1984) and others who have pointed to the unnecessarily restrictive policy pursued in the eighties in relation to the PSBR. We recognize the need both for public investment programmes and for a greater stimulus to private investment. Such investment is essential to broaden the capital stock and to diffuse the new technologies. However, the importance of the structural changes taking place and still required in the UK economy to realize the scope for employment growth need to be recognized to a far greater extent. In this chapter we illustrate this point with respect to technology policy, education and training policy, regional policy and incomes policy.

The detailed analysis of the scope and potential of information technology attempted in chapters 4 and 10 brought to the forefront the relatively radical 'paradigm' or 'technological regime' change of the new technology. Its main characteristics are very briefly:

1 A significant increased potential for flexibility both in manufacturing and in services, opening up an increasing number of service activities to trade.

2 A potential for productivity growth in relation to all factor inputs, ranging from labour, capital, materials and energy, to the other more intangible factor inputs such as inventory stock control, organization, quality control, management, marketing, etc.
3 A potential for decentralized, yet more interactive decision making and organization.

All three features indicate the need for widespread organizational and institutional change. This need is as obvious at the level of the firm and technology policy, which we discuss in the first sections of this chapter, as it is at the level of work organization, education and training which we discuss next, or of other government policies, both national and regional, which we discuss in the last two sections.

Organizational Change Within the Firm

Before discussing any specific policy implications, we start this section with some elaboration of the most distinctive features of IT as they emerged from the two IT chapters, focusing in particular on what this might imply in terms of the required need for organizational change within the firm. An excellent and comprehensive discussion of this issue can be found in Perez (1985), from which we retain the following main points.

Flexibility

The further application of IT will accentuate and in some cases increase dramatically the flexibility of the firm, both at the level of the production process and at the level of the design of products.

At the level of the production process this implies a reduction of the importance of plant economies of scale – a typical characteristic of the old 'technological paradigm' – with the increased possibility of more rapid 'retooling' through flexible manufacturing systems. These points came out very clearly in the MIT study of the world automobile industry which pointed to the enormous productivity gains achieved by Japanese firms through the combination of 'flexible automation' and the 'Just-in-Time' system (Jones, 1985). As Perez (1985, p. 21) points out:

When production – as well as productivity – depends on the repetitive movements of motors and workers and every change of model or tooling is downtime, optimal production costs are closely related to achieving high volume production of identical units. With electronic controls and the relatively low cost of programming rapid changes in production schedules, such limitations disappear to a great extent. It is of course still possible to apply the new technologies for mass production of certain components or products, at a scale which could be a multiple of the previously established optimal size. However, the most significant change, rich in eventual combinations, is a quantum jump in potential productivity for small and medium batch production.

At the level of the product, the over-riding need for 'mass production' is also put directly into question. The application of IT computer-aided design and other IT developments will often allow for more rapid and less expensive changes in design and a more differentiated range of final products. Again this is clearly observable in the case of the motor vehicles industry. This opens up possibilities of capturing and generating new, income elastic, demand with less emphasis on price competitiveness, and more on quality and specialized features.

In chapter 10, it was argued that this would also open up many service activities to 'trade', bringing about employment growth in many of the IT-producing and -using services and possibly even lifting the UK's manufacture balance of payments constraint on future economic recovery growth.

Pervasive Productivity Growth Potential

As we emphasized in chapters 4 and 10, IT with its potential to collect, generate, analyse and diffuse enormous quantities of information at minimal costs affects the productivity of all factor inputs. It is also this feature which distinguishes IT so clearly from 'old-fashioned' automation. Some of the most significant productivity gains linked to the introduction of IT relate to more efficient inventory control, as well as energy, materials and capital savings.

In so far as most firms, as well as government itself, despite recent trends in wage costs, still appear primarily if not solely motivated by the need to keep labour costs down in their IT investment decisions, the overall productivity growth potential of the more widespread application of IT has barely been realized. For most firms the introduction of IT is still viewed as a more or less straightforward, undoubtedly more sophisticated form of automation.

Again we take up below the policy implications of this often unperceived 'pervasive' productivity potential.

Organizational Change

The application of IT questions virtually everywhere the traditional organizational forms of decision-making within the firm, as well as within practically any other form of organization. At the production level the application of IT creates possibilities for decentralization, as much geographically as organizationally. Assembly-line production, with its hierarchical professional management organizational form of decision making no longer corresponds from such a perspective to the optimal technological mode of organization. The previous mode of organization, in Perez's words

implied a sharp separation of plant management from economic management and, within each, a clear differentiation of activities in order to identify all forms of repetitiveness for subsequent automation. It was mainly an analytical mode. It demanded focusing attention on parts or elements of processes; it led to detailed definition of tasks, posts, departments, sections, responsibilities and to complex hierarchies.

By contrast, the application of IT could be said to imply a more 'synthetic' organizational form of decision making. It focuses on a more interactive form of organization in which both managerial and productive activities are becoming integrated, as we saw in the case of the 'Cultural Revolution' at General Motors (Gooding, 1984). The concept of 'systemation' suggested by Perez (1985, p. 30) is probably the best description of what is meant. As she puts it:

In a sense, it could be said that information technology does for the firm what the assembly line did for the plant. The firm, as a whole, becomes a continuous flow system of activities, information, evaluations and decisions. But there is a crucial difference: whereas the assembly line was based on the constant repetition of the same sequence of movements, information technology is based on a system of feedback loops for the optimisation of the most diverse and – changing – activities.

The implications of IT for organizational change, however, go well beyond the firm as an organizational entity. The vast array of government administration, research organizations, social and educational organizations and many more will be, in a similar way, subject to radical change as the result of the application of IT.

We now turn to a brief summary of the main policy implications of these features of IT.

Implications for Technology Policy

All three factors point to the radical change potential associated with the adoption of IT. They explain, in our view, both the slowness with which firms appear to have adopted these radically more efficient technologies and the sometimes disappointing realization of the potential productivity gains. The historical analogy with electricity made in chapter 4 is also instructive here. This analysis points to the persistent need for a specific technological dimension in the formulation of economic policy.

Hitherto in the United Kingdom and most other OECD countries, technology policies have been considered as something rather separate from economic policy, or as relevant mainly for defence or energy policy. One of the reasons for this is undoubtedly the almost exclusive focus of technology policies on R & D and innovation. The question as to the diffusion of innovations has hardly been addressed. As Paul David (1982) puts it: 'Innovation has . . . become our cherished child, doted upon by all concerned with maintaining competitiveness and the renewal of failing industries; whereas diffusion has fallen into the woeful role of Cinderella, a drudge-like creature who tends to be overlooked when the summons arrives to attend the Technology Policy Ball.'

Leaving aside the case of agriculture (where diffusion has been a major policy concern for a long time with great practical benefits), the present levels of government support speak for themselves. In the United States, for example, of the nearly $17 billion of federally supported R & D in 1974, a mere $43 million was directed to 'secondary utilization of the technologies developed'. No comparable figures are available for the United Kingdom but there is some reason to suppose that proportionately they are a little higher through such schemes as the microelectronic awareness programme. Nevertheless the general point is also valid for the United Kingdom. One of the main reasons for this neglect can be found in the widely held view that 'diffusion' assistance would amount to unfair and unwise interference with the market system. R & D government support by contrast has both economic theory (since Arrow's (1962) seminal information paradox article) and economic policy credentials (see, e.g., the national security arguments in relation to military R & D or the patent system and the need for intellectual property protection). But the arguments against an active diffusion policy and for treating R & D very differently are not really convincing.

First, as Rosenberg (1975, 1982) in particular has illustrated, whereas the analytical distinction between innovation and diffusion might at first sight appear a useful one, and indeed is so for many purposes, it does not imply that the two concepts can really be completely separated. Typically with the further diffusion of new technologies, continuous improvements will occur to the original innovation. As Kline and Rosenberg (1985) put it:

It is a serious mistake to treat an innovation as if it were a well-defined, homogeneous thing that could be identified as entering the economy at a precise date. . . . The fact is that most important innovations go through rather drastic changes over their lifetimes – changes that may, and often do, totally transform their economic significance. The subsequent improvements in an invention after its first introduction may be vastly more important, economically, than the initial availability of the invention in its original form . . . consider the performance characteristics of the telephone around 1880, the automobile, vintage 1900, or the airplane when the Wright Brothers achieved their first heavier-than-air flight in 1903 – in that form, at best a frail and economically worthless novelty.

In other words, diffusion, with its significant and continuous flow of incremental and improvement innovations based on user feedback effects, is an integral part of the process of technical change. As a consequence, technology policies geared towards R & D and innovation should also affect directly the diffusion process. Similarly, policies directed towards the diffusion of innovations (one can think of MITI and its role in the transfer of foreign technology to Japan) will also affect actual innovation and R & D, and indeed should be designed to do so.

Secondly, and more directly related to our present concern, when confronted with a generic, pervasive set of inter-related innovations such as IT, the realization of the productivity growth potential will even more be based on the close interaction between improvement innovations and diffusion within the specific user environment, so-called 'learning-by-using'. The rapid growth of software services is a typical reflection of this further 'systemic' integration between R & D, incremental innovation, learning-by-using and diffusion.

Thirdly, the long diffusion lags and in some cases limited spill-over effects both reflect the technical characteristics of the innovation and the characteristics of the user and supplying firm and industry, as well as the overall macroeconomic environment. (In cases of radical new technologies such as IT the learning by using and adaptation of the innovation to the specific user's environment might be a long and painstaking process.) In contrast to many countries' R & D and innovation policy, a more systemic technology policy aimed also at diffusion offers consequently plenty of scope for integration into overall macroeconomic policy. The micro-electronic awareness and development programmes promoted by the Department of Industry are an interesting and at least partly successful example.

Fourthly, and on a narrower basis, with the heavy and in some countries increasing support for military R & D, a clear case for a more active 'diffusion policy' geared towards the stimulus of civilian spill-over effects can be made. With the accentuation of the arms race, there is a clear problem of growing inventories of unused research results. The setting up of specific 'information' programmes for private, civilian firms from military R & D is a first step in the right direction, but insufficient in itself. Especially in the United Kingdom, there is also a need to make available a part of the capacity of the defence R & D system for civil applications.

Fifthly, particularly with regard to foreign technology, diffusion policies might have a particular role to play. For industries which are at some distance from the world technological frontier, such a policy seems extremely relevant. But even for technological leaders, active support in seeking and using the best available world technology is common sense. No country can be a technological leader in all areas, and all can learn from international experience. However, it is the leading-edge civil technologies which require the most intense efforts. This points in particular to the importance of technology transfer from Japan. The experience of the leading British computer firm, ICL (now part of STC), in its collaboration with the Japanese firm Fujitsu, demonstrates the great advantages of technical collaboration of this type.

What these diffusion policies might imply in practice is again very much dependent on the specific characteristics and the existing institutional framework of the country. In some countries, such as Japan, this might represent little more than a continuation of existing policies. In the case of the United Kingdom it would represent a more radical shift from a preoccupation with 'pioneering technological frontiers' to a more 'pedestrian' support for the adoption and incremental improvement along the diffusion path of 'existing' technologies in all branches of manufacturing and services. The main thrust of government support for R & D in the United Kingdom has hitherto represented a serious misallocation of resources, since it was directed to areas and projects which were very unlikely to yield economic benefits. Much of it went to very large aircraft and nuclear projects and to military projects. The incorporation of a more explicit economics dimension in this area of decision making could help to rectify and prevent this misallocation. The money spent on R & D may often be wasted if the production and marketing follow-through is lacking.

Since the seventies technology policy in the United Kingdom has begun to move a little in the directions we have indicated. Schemes to stimulate awareness and applications of microelectronic technology were introduced already then by the Department of Industry. More recently the 'Alvey programme' was launched in 1983 in response to the Japanese announcement of a programme to promote the development of a new generation of computers. It involves the expenditure of £350 million by government and industry on 'precommercial R & D' in four main areas of technology relevant to '5th Generation' computing – VLSI (very large-scale integrated circuits). MMI (man–machine interface), software engineering and IKBS (intelligent knowledge-based systems).

The Alvey programme is a notable advance in several respects. First, it represents a major commitment to the promotion of 'generic' technologies although of course the programme involves the support of hundreds of projects. Secondly, it is based in principle on collaboration between universities and firms and on 'clubs' of participants. Thirdly, it involves some attempts to identify strategically important areas of development work within the IT paradigm. Discussions are now in progress (1986) on a possible follow-up – the so-called 'After Alvey' programme, and the Report of the 'IT 86' (Bide) Committee. This Report puts a welcome stress on user involvement in the next stage.

These existing initiatives provide an important base on which to build a more comprehensive UK technology policy, with a greater emphasis on diffusion of the new techno-economic paradigm, as well as on the further development of the key technologies. Although it has several very welcome and innovative features the first Alvey programme was too narrow in both scope and objectives, and too limited in scale. In the follow-up much greater emphasis needs to be placed on the participation of potential users of the new systems and their problems, and on future home and export markets. The demonstration projects of the Alvey programme are a step in the right direction, but this principle needs to be carried much further. There is now abundant evidence that a combination of demonstration and user involvement is essential for effective and rapid diffusion of new technologies. It is also essential for the humanization of this technology.

Much greater consideration also needs to be given to the links between R & D policy and industrial strategy more generally. The UK electronics industry has suffered from a one-sided orientation towards exotic (and costly) defence R & D and an inadequate involvement in large-scale civil markets in consumer electronics and capital goods for civil applications. It has lagged very seriously in CNC tools and robotics and in the ability to deliver high quality components for civil applications.

One of many indications of this critical weakness came from IBM (UK) which warned its UK suppliers in 1985 (Bird, 1985). The Chairman of IBM (UK) said that:

UK sources are becoming a serious concern. Our primary concern relates chiefly to the technological needs we foresee we will need in the coming years in ever greater quantities. This includes magnetic and optical storage, printing technologies, display devices and communications. While the UK does have a presence in some of these technologies, much of it is bound up in aerospace, communications and defence, with little or no native capacity in the commercial sector.

Firms such as IBM and Marks and Spencers, in their different areas of activity, can and do play a key role in raising technical standards and improving quality through their procurement policies in relation to local suppliers. But their efforts need to be underpinned by a government strategy, which also uses the immense government procurement power for these ends, but more importantly strengthens the capacity of the industry as a whole to meet the future requirements of the world civil markets. This means an industrial policy linking R & D with investment, procurement, marketing and industrial restructuring.

The level of British R & D itself does however need to be strengthened within an overall industrial policy in which the main stress is on diffusion and applications of new technology. The days when Britain could be regarded as high in the 'League Table' of industrial R & D have long since gone. The United Kingdom now spends much less on civil R & D as a ratio of GNP than Japan, Germany, Sweden or Switzerland, and whereas these countries have been increasing both their public and private outlays in the eighties, UK expenditures have stagnated or declined in real terms, except in the defence area. A much more ambitious 'After Alvey' programme, combined with other similar programmes in areas such as biotechnology and materials technology, would be needed to halt this relative decline, as well as other initiatives.

Nor can fundamental research be neglected. One reason that the Alvey programme (and similar programmes in other countries) put such stress on university–industry links is that the interaction between scientific research and industrial technology is now much more intimate in the new technologies. Our proposals should in no way be construed as a down-grading of basic or applied research. These certainly need to be strengthened considerably but within the overall context of an industrial policy directed to the efficient diffusion and exploitation of new technology.

Changes in Occupational and Work Organization

Directly related to the increased flexibility potential of the application of IT, at both the production process and product levels, is the increased need for flexibility in the organization of work, for multiple skills, for retraining and more generally for computer literacy. These needs are as urgent at management level as they are at blue collar level.

In terms of work organization, some countries have already gone far in removing the various institutional barriers to allow for more flexible work patterns, work sharing and part-time work. In most countries though, rigid fixed work-time patterns are the norm and various forms of financial or career discrimination exist against those opting for more flexible work-time organization, which reinforce 'natural' conservatism. As Marie

Jahoda has argued in chapter 1, various part-time work arrangements for working mothers (and fathers) and for other people are to be preferred to early retirement or withdrawal from the work force for both social and economic reasons.

At the occupational level, the widespread application of IT is blurring many occupational categories. Apart from the specific IT skills, the increased need for multiple skills and the change in content of specific occupations will further reinforce the need for regular retraining and occupation switching. This is most evident in industries such as vehicles and printing, where traditional craft boundaries have limited and retarded the potential productivity gains from new technology.

From a policy perspective this points to the need for:

1 A greatly increased supply of highly qualified people in computer-related technologies, electronics and system designs.
2 A more highly educated work force – at all levels of the occupational ladder – capable of flexible adaptation to technical change, learning a variety of new skills during the working life.
3 A major drive to computer literacy throughout education (the fourth r).
4 Thorough systems of consultation with the work force in relation to technical change.

These points are closely inter-related. The drive for 'computer literacy' in the schools cannot succeed without a more professional and better qualified group of teachers. The shortages of highly skilled people in industry and the low priority given to education in the eighties have attracted some of the best teachers away from the universities and schools. This point was very clearly brought out in a survey of computers in schools in the USA by the *Wall Street Journal* in April 1985:

The more than one million computers in US public schools are supposed to be changing the shape of American education. . . . But the rush to buy classroom computers – the number has increased 60% this school year – doesn't seem to be doing the youngsters using them much good. At least not yet. . . . One major obstacle is a shortage of teachers who understand both their regular subjects and fast-changing technology. . . . What many schools do when the computers are coming is to simply draft gurus – designating one or more teachers as instant computer experts. . . . And, after computer gurus do become knowledgeable, they may feel so overworked and underappreciated that they leave the school system.

Similar and even more acute problems certainly exist in British schools and other schools in Europe.

Implications for Education and Training Policy

Indeed, this point about the 'seed-corn' for future generations in the education system is probably one of the most important areas of universal agreement amongst studies in this area, which in the United Kingdom even more than in the United States repeatedly point to skill bottlenecks as the main obstacle to innovation and adaptation. This universal agreement is however in sharp contrast with some of the macro evidence traditionally used by labour economists. That evidence based on aggregate numbers of unemployed compared with vacancies tends to suggest that structural, so-called occupational 'mismatch' unemployment is minimal (see Layard et al. (1984). Such evidence will in our view give a completely misleading picture, since the 'vacancies' have an importance out of all proportion to their numbers, and the retraining issue is completely obscured. The issue has come to the forefront in the debate which followed the Finniston Report (1980) calling for more technical engineering skills in the UK. This report because of its unequivocal request for more engineers ('there can be no danger of producing too many engineers') has been dismissed 'as polemic on behalf of the engineering lobby (Mace, 1980); the argument being that engineers can be substituted for (e.g. technicians) and that pay levels do not reflect any engineering skill shortages. However, as Senker (1986) puts it forcefully:

In Britain, the economists' case against training more professional engineers is that there are no indications (e.g. in terms of rising relative salaries) that engineers are in short supply. But for these economists, firms' strategies are not variable. They are fixed. Entrepreneurs buy factors of production in the cheapest markets, mix them together in the optimum combination to maximise their profits in (given) product markets. In a real sense, the entrepreneurs of the theory have no discretion – they are automata. They cannot decide to go up market, to go down market in terms of the products they manufacture. They cannot decide to improve their production processes, or not to do so. This does not reflect the key decisions facing firms, and the results of such analysis are liable to mislead policymakers. (Senker, 1986, p. 17)

The point made by Senker (see also Brady and Senker, 1986) is well taken. Long-term shortage of skilled people will lead firms to adopt their product and process strategy to lower levels of achievement and growth, reinforcing the vicious circle of long-term relative decline of the international competitiveness of the UK economy.

This points to the need for both specific and general measures to improve the quality and quantity of educated and trained people in the United Kingdom. There has of course been a major expansion of expenditure on the MSC and especially on the Youth Training Scheme, but even more needs to be done especially in the field of adult training. Numerous studies have pointed repeatedly to the comparative weakness of the United Kingdom in this area by comparison with the German Federal Republic, Japan and Sweden. It has often been a case of too little and too late in the reforms designed to catch up, with disastrous consequences for the competitive performance of the country. The low priority accorded to education in the eighties and the uncertainty about the future of the Training Boards have exacerbated an already bad situation.

Specifically, there needs to be a major effort directed to the supply of the key IT skills. This could perhaps best be organized as a flexible operation by a special Directorate (comparable to the Alvey Directorate but working in the sphere of education and training), with the resources to initiate, sustain, subsidize and otherwise encourage those initiatives in all types of educational institutions needed to overcome the immediate skill shortage problems and to improve the supply of the key skills involved. The nurturing of the 'seed corn' in both schools and universities requires special financial measures on the lines of the NSF programme for American universities.

More generally, there needs to be an expansion of the education and training system, involving especially flexible postexperience courses in higher education, organized in close cooperation with professional organizations, local and regional authorities and organizations. There needs also to be greater encouragement and financial support for school children to continue their studies and/or vocational training. There will be an increasing need for a generally well-educated population, as well as for continued and flexible training by and for industry.

There is now a strong case for policies designed to persuade more young people in the 16–19 years age bracket to stay on at school. Compared with their contemporaries who receive training allowances or supplementary benefits, the young people who stay on at school get nothing, other than child benefit. The expansion of the Youth Training Scheme, welcome though it is, has thus acted as an incentive, together with the decline in the education system, for young people to leave school. This trend needs to be reversed together with a reversal of the decline in higher education. The universities and polytechnics should have a key role in postexperience education for all professions.

There is now also an opportunity to improve standards throughout the education system and to raise morale in the profession by reducing class size in the school system. All of this would imply a considerable increase in expenditure on both education and training and this is indeed what we propose, since we regard an improvement in this area as the single most important contribution which government policy can make to the efficient adoption of new technology and the reduction of structural unemployment.

The evidence is strong that the unskilled and uneducated are experiencing increasing difficulties in the labour market in all industrial countries. Educational and skill enhancement is the main way to tackle this problem rather than wage reduction. Education and training are not simply consumption to be cut back in difficult times. They are the most important long-term investment for the future.

Institutional Changes: Regional Policy and Telecommunications

IT represents a formidable challenge in terms of the required need for radical institutional change. Most of our present institutions were created under and are still geared towards the older technological paradigm. These institutions, with their self-perpetuating interest groups, represent probably today the most formidable barriers to the rapid diffusion of IT and the realization of its potential growth and productivity gains. With regard to the United Kingdom the list is particularly vast and diverse. For example, the role played by the low rent housing market in reinforcing geographical labour immobility is probably an even more severe barrier than the existence of a 'craft' and 'trades' union structure with respect to occupational flexibility. A major public housing programme would contribute both indirectly and directly to the improvement of economic performance and the reduction of unemployment. In the public sector there is a need to develop a computerized information system to facilitate the exchange of council houses, which at the moment is haphazard and very difficult, but to increase mobility it would be essential to promote much greater flexibility both in the types of accommodation and the arrangements for exchange and transfer. Especially important would be local authority housing programmes geared specifically to the needs of young single people and young couples.

In order not to get lost in an enumeration of these institutional bottle-necks, we limit our discussion here to the required need for institutional change in two specific areas of rather widespread (mis)use in the United Kingdom and many other European countries: regional expansionary policies and the national telecommunications monopolies.

Regional expansionary policies were traditionally aimed at distributing on a geographically more widespread basis the employment growth effects of the postwar growth boom. These policies have also led to the creation of new towns and mass-transit infrastructural investment. They often had strong political regionalistic undertones. In practice though most of these policies consisted in the provision of various investment incentives. Their direct aim was the regional creation of employment opportunities, not necessarily through the net creation of new jobs but through the geographical redistribution of employment opportunities within the country. In the fifties and sixties they were largely effective from this point of view, probably more from the multiplier effects of new investment than from direct job creation. The regional distribution of unemployment was far more uneven in the United Kingdom in the thirties than in the eighties.

However, it is probably also fair to say that, viewed in retrospect, such policies also had a 'boomerang' effect. By artificially cheapening investment cost, they provided incentives to attract more capital-intensive industries and in the long run may have led to a further accentuation in the capital-using labour-saving direction of technical change. Combined with the tendency to tax employment, through employer's social security contributions, these policies could be said to have accelerated the emerging 'capital shortage' crisis of the eighties, typical of the United Kingdom and many European countries and might well have reduced the overall employment creation potential of the economy.

Such policies are now clearly inappropriate. In the United Kingdom, the present government has already dismantled some of the regional investment subsidy systems, partly because of their adverse long-term effects. However, the dismantling of by and large inappropriate policies will not solve or 'dismantle' the problem of high regional unemployment and low growth. If anything the continuous process of structural change in the economy will exacerbate problems of regional decline. Again it could be argued that IT provides some distinctive features upon which the regional policy issue could be focused. As in the case of Northern Finland (Oulu) and Central Scotland ('Silicon Glen') this points to the great importance of educational and technical infrastructure, as well as good communications and environmental policies. Others are more directly related to the unique information gathering and distributing potential of the new technology. The Scottish Development Agency is one example of the benefits which can be achieved by determined efforts to build a technological dimension into regional policy.

It is particularly notable that in Japan, which has had a rather low rate of unemployment in the seventies and eighties, new government policies were adopted in the early eighties to cope with the problem of uneven regional development. These policies were specifically intended to enable those prefectures which were suffering from the decline of the old so-called 'smoke stack industries' to benefit from the diffusion of new technologies. The studies which led the government to adopt a flexible programme of tax incentives, public investment and other measures, emphasized that the key elements in generating new employment in these regions were infrastructural investment in research, education and communications (Kuwahara, 1984).

Limiting ourselves to a few rather broad and even futuristic points of analysis, we would stress in particular the following ways in which IT will affect regional policy.

First, the dramatic potential offered through IT to separate actual production from control, administration, design, management and marketing. This offers substantial scope for the geographical relocation of firms and businesses. In some occupations it opens up the possibility of work at home.

Second, the realization of this potential depends crucially on the widespread availability of a fully interactive (two-way) communications network.

Third, the decentralization potential of IT, will allow in a far more effective way for the creation of local, quasi-autonomous growth poles. Just as in the case of 'economies of scale', IT questions the basis itself of the agglomeration effects, generally associated with growth concentration.

Fourth, and following to some extent from the previous points, in the long run, IT could ultimately lead to the reduced importance of daily 'commuter' type transport services.

All four aspects have important implications for regional development and regional policy. In the sphere of technology they point to the crucial role played by the telecommunications sector, in 'facilitating' this regional growth potential. In the United Kingdom they seem to suggest a need for the strengthening of the regional dimension in local government, but this is beyond the scope of this book. Suffice it to say that the strengthening of the capacity of local authorities to initiate and contribute financially to new developments with employment implications, as has already occurred here in a small way with the 'Enterprise Boards', is an essential complement to any central government policy. The contrast here with the situation in countries such as the German Federal Republic and the United States is very striking.

The generally restrictive attitude of central government towards local authority finance in the eighties, as in the reduction of the rate support grant, has gravely limited the ability of local governments to promote new initiatives and employment generating activities. Despite this, a number of local authorities have taken useful steps, as for example in the setting up of 'enterprise workshops' to facilitate the launch and success of new small firms and cooperatives in Sheffield and other similar initiatives elsewhere.

Because of its crucial importance both for regional development and for the diffusion of the new technology paradigm more generally, the second broad and general example of institutional change which we wish to comment upon briefly is another hotly debated area, the widespread existence of national telecommunications monopolies in most OECD countries (at least until 1983). Infrastructural monopolies from electricity generation and railroads to telecommunications have traditionally been justified on technical, economies-of-scale, grounds. Where these were more regional in focus, regional infrastructural monopolies were set up. In relation to telecommunications, the point of debate has been that precisely with the advent of more flexible IT equipment, such technical economies of scale arguments are becoming somewhat less important, and regional and local flexibility have become more important issues.

From an economic point of view, monopoly power can be identified with advantages in the scale of R & D, learning-by-doing, and in the tele-communications case more integrated telephone equipment manufacturing and servicing. At the same time, though, apart from the obvious lack of price competitiveness, inefficiencies or delays in innovation and limited consumer product choice have also often been identified with telecommunications monopoly power and with oligopolistic procurement arrangements. The massive documentation on the AT & T hearings and the experiences of the 'Ring' in Britain provide ample information on this subject. In our view, and limiting ourselves exclusively to the technology side of the issue, it is clear that monopolists – whether state or private – may often delay the introduction of new products or services, and may be slow to react to the new demand opportunities created by the new technology (Englander, 1985). The contrast as illustrated in chapter 9 between the slow growth in output in the telecommunications sector and the innumerable new growth openings both in products and services is indeed striking.

The development of independent local area networks and new sources of supply as well as the end of the institutional telecommunications infrastructural monopoly are also an essential part of diffusion policy. The diffusion of telecommunications technology despite its hundred year innovation credentials is still in its infant stages. There are still plenty of demand niches which need to be identified, where the long, painstaking process of incremental change and user-feedback effects have yet to be set in motion. In view of the generic nature of the technology, it would be surprising if one single firm or a few equipment suppliers were able to cope with the enormous diversity in user demand and requirements.

An extremely important related issue is the provision of a wide band-width cable network as infrastructure for many new types of information service. Public policies and standards and, in some cases, public infrastructural investment are inevitably involved here. Once again, Japanese policies for infrastructural investment are far more ambitious (Arnold and Guy, 1986) and Mackintosh (1986) has made a convincing case for a very ambitious EEC initiative in this field.

Wage and Income Policies

Although it clearly emerges as an important issue from the factor substitution discussion in chapter 3, we have so far in this chapter paid scant attention to the issue of wage and income policies which is a major focus for both neo-classical and Keynesian analysis of unemployment. Today, as in the heyday of classical economics (see, e.g., our quote from Von Mises in chapter 2) the present levels of unemployment have often been identified with 'wage gaps': the gap between the actual wage and the 'full employment' wage (for recent empirical evidence see Bruno and Sachs (1985)). Until recently, the empirical evidence did not generally support the expected negative relationship between employment growth and real wages and it is still equivocal with various country studies pointing in different directions.

It is outside the scope of this book to add anything to this empirical debate, but we note that in the debate about the relative importance of aggregate demand as opposed to real factor prices, the recent evidence in favour of real factor prices remains very much country specific. One of the reasons is probably the difference in the institutional setting in the various countries with regard to unionization, union militancy, minimum wage laws, incomes policy and wage bargaining. The development of social mechanisms to constrain the rate of growth of real incomes in line with productivity is undoubtedly an essential part of any effective counter-inflationary macroeconomic policy and to that extent we agree with Layard and Nickell (1985) that incomes policies can be thought of as ways of reducing the 'natural' rate of unemployment and have from this perspective (at least in the United Kingdom) been unjustly criticized as ineffective and unworkable. Probably they will be most effective if based on mutual agreement between employees and unions at national level, as in Sweden.

However, as we explained in chapter 3, we do not consider attempts to generate a reversal of labour-saving trends by temporary reductions in the relative price of labour as effective or desirable. First, as illustrated in the so-called efficiency wage literature, there is nothing illogical in firms paying out higher wages once training and exit costs are taken into account. As Stiglitz puts it:

Thus, unemployment is caused by too high wages, but the excessively high wages come about not because of union pressure; rather, the interactions of the wage policies of the different firms, and the inability of the different firms to coordinate their wage policies, lead to an inefficient Nash equilibrium. (p. 607)

In the case of such undoubtedly more realistic wage formation models, the simple emergence and further diffusion of a new technology in its stylized neoclassical form: embodied and purely labour augmenting, will bring the macroeconomic full employment wage down, without any incentive for the individual firm to do so. Unemployment will follow. Whether it is in this case the inherent downward wage inflexibility or the new technology which will be the cause of the increase in 'classical' unemployment is in our view no more than a semantic issue. What is clear though is the fact that real or nominal wage rigidity is inherently part of the system.[1]

Second and with respect to the traditionally invoked factor price inducement mechanisms on the rate and direction of technical change, it will be clear from our discussion in chapter 3, that only a reversal of long-term expectations and/or a sudden drastic reduction in real wages would be likely to prove effective. In so far as short-term changes affect innovative behaviour, we would agree with Rosenberg (1982) that sudden big changes in factor availability are much more significant than small changes in relative factor prices. We would also accept much of Rosenberg's insistence on the importance of other inducement mechanisms for innovative activities in addition to factor costs, and in particular his point that there are inducement mechanisms operating in relation to particular specific skilled labour bottlenecks, for both economic and social reasons.

In R & D and innovative activities generally, long-term secular trends have indeed a much greater influence than short-term fluctuations. As we emphasized in chapter 3 there are 'natural trajectories' with any technology which indicate the possibility of labour-saving improvements through standardization of components and the substitution of machinery for labour as a technology matures over a long period. However, whereas entrepreneurs and engineers welcome all types of cost-saving innovations, we would accept the inherent plausibility of a long-term labour-saving bias in the direction of technical change as postulated by Hicks.

Even with respect to IT, despite the fact that it would offer more scope for faster adaptation to short-term factor price changes, we would still expect any factor price-induced effects to show up with significant lags. Thus, whereas over the last couple of years real wages have sometimes fallen and real interest rates have been at historical heights, we would expect the capital-saving potential of IT to diffuse relatively slowly. A similar point could be made with regard to the recent decline in the nominal and real price of oil and the nevertheless continuing trend towards energy-saving investment, both at the firm level and the individual consumer level (cars, house insulation, etc).

This is where the concept of 'paradigm' change, of change in 'technological regime' is in our view so enriching. These changes in managerial and engineering 'common sense' go well beyond the short-term factor price adjustments that economic theory has generally focused upon. They involve widespread and sometimes radical organizational and institutional change and savings in all factors of production, as we tried to illustrate in some of the previous policy discussions.

This implies that although, in order to achieve sustained employment growth, we would see some scope for macroeconomic wage and incomes policies, from a technological and particularly from an IT perspective we would see more scope for long-term occupational wage flexibility, and occupational mobility, with the major training implications that this implies.

The most important condition for increasing employment is sustained increase in investment and in productivity. As employment rises and unemployment diminishes, real wages can certainly increase without damaging profits, provided productivity increase is maintained. It is for this reason that we have stressed so strongly the importance of diffusion of new technology and of skills.

Paradoxically, whereas a traditional neo-classical analysis would present today a relatively bleak empirical assessment of the employment compensation potential of new technologies, given the low price and substitution elasticities in most industries, our own analysis would come up with a more optimistic compensation scenario pointing in the first instance to the employment growth opportunities in the new technology industries and products both in terms of the domestic market and in terms of the export market. However, the potential for compensation would be crucially dependent on the supply of capital and new investments, and on the development of an appropriate institutional framework to give the new technologies full scope and to retain a competitive position in international trade.

It is already evident that this is no mere hair-splitting academic debate. As we emphasized in our earlier book, the policy implications were very clearly brought out in the debates which took place in Japan soon after the Second World War. One of the few British economists to study these developments was G. C. Allen (1981). According to his account the Bank of Japan at that time offered orthodox 'Western' style economic advice to the Japanese Government, based on the view that Japan should concentrate on industries where low labour costs would confer a comparative advantage. But it was not long before the Government began to disregard this essentially static advice and follow the long-term dynamic strategies advocated by the Ministry of International Trade and Industry (MITI) for the encouragement of the growth of new industries (without any special regard for relative factor costs). Discussing the advice tendered by MITI, Allen observes:

Some of these advisers were engineers who had been drawn by the war into the management of public affairs. They were the last people to allow themselves to be guided by the half-light of economic theory. Their instinct was to find a solution for Japan's postwar difficulties on the supply side, in enhanced technical efficiency and innovations in production. They thought in dynamic terms. Their policies were designed to furnish the drive and to raise the finance for an economy that might be created rather than simply to make the best use of the resources it then possessed.

In our view it could be dangerous to focus attention exclusively on the best use of the 'resources we now possess' and to neglect the 'economy that might be created'. It is for this reason that we have concentrated on policies for the new technologies in the economy we might create in the United Kingdom.

Note

1 As Stiglitz put it: 'it is easy to see in this model how a disturbance to the economy can move it from a position of full employment to one of unemployment. Assume, initially, that the economy were at full employment but then the technology is disturbed in such a way as to decrease the marginal product of labour. So long as all other firms continue to pay w∗, it pays each firm to continue to do so: all adjustments take place in the number of workers hired. If all firms believe that other firms will keep nominal wages constant, in spite of a change in prices, it pays each firm to keep its wages constant in nominal terms: the model is consistent with either nominal or real wage rigidity.' (1985, p. 607)

Conclusions

Christopher Freeman and Luc Soete

In chapter 11 we made some tentative estimates of the new employment which might be generated in the UK economy over the next 5–10 years on the assumption of some major changes in policy with respect to public investment, the approach to education, training, research, development, diffusion of new technology and technology policies generally. In chapter 12 we have discussed the broad outline of such policies.

It is not possible in this book to prescribe detailed recommendations in all these areas. We have confined ourselves to indicating the broad direction of the policy changes which are necessary. The translation of these broad indications into policy formulation through legislation, and various specific economic and technology policy instruments must be a matter for those political decision makers (if any) and their advisors who might accept the type of analysis which we have made.

We would stress in conclusion that what we are advocating will not be easy for any government or any party to accept. But we believe that only some fairly decisive changes on the lines we are suggesting have any chance of halting the long-term relative decline of the British economy, which has so far been temporarily masked by the North Sea oil revenues.

Since the Second World War the British economy has repeatedly suffered from a balance of payments constraint to more rapid economic growth, from slow productivity gains, from skill shortages engendered by even a moderate expansion, from inadequate professional management, from poor industrial relations, and from serious misallocation of resources in R & D and technology generally. We have argued that it also suffers, and will increasingly suffer if expansion is resumed on a sufficient scale, from a capital mismatch problem. Profound structural changes are essential to overcome these problems and to embark on the sustained growth necessary to achieve high levels of employment.

We do believe that full employment is an attainable goal but we are only too well aware that it will not be easy for any government to change the long-term constraints. Perhaps the nearest thing to what we have in mind was Harold Wilson's 'white-hot technological revolution', but this turned out to be a nine-months' wonder rather than an effort sustained over decades to change the level of technological competence throughout the British economy.

Even if major changes in policies for technology were adopted on the lines we have suggested in chapter 12, it would still be a matter of years, rather than months, before the effects of such policies could be felt. To generate a great deal of new employment therefore, requires a combination of short-, medium- and long-term policies.

Chapters 7 and 8 have illustrated some of the problems of generating new employment in the manufacturing sector and as they show, the only manufacturing sectors registering employment gains in the past few years have been those closely related to microelectronic technology, such as computers and electronic components. A sustained investment boom could generate new manufacturing employment capacity in these sectors, as occurred in major re-equipment booms in the past, but we cannot look to manufacturing or the 'network services' as a major sector of employment gain over the next 5 years.

A reduction in unemployment, therefore, depends almost certainly on other sectors of the economy. Whilst no significant employment growth can be expected in the primary sector over the next 10 years, the growth of service employment could be by far the most important factor. But as chapters 8, 9 and 10 have shown, the interdependencies between manufacturing and services are very great and the boundaries are becoming blurred. Further growth in service employment depends heavily on maintaining an internationally competitive manufacturing sector. It also depends increasingly on technical change in the service sector itself. For this reason the rapid and efficient diffusion of information technology is vital for the future of employment growth throughout the economy. This does not mean, however, that the new jobs will be mainly in the IT-producer industries themselves. On the contrary, in the short or medium term, only a small proportion of the total employment will be directly in the IT-producer industries and services although a higher proportion will be in IT occupations and in IT-user industries.

As chapter 9 explains in detail, some special features of the service industries must be taken into account, when considering the potential for further employment growth. In the first place, the most labour-intensive areas are mainly direct public service employment (central, regional and local government), or are heavily dependent on public funding (education and health). In most OECD countries and especially in the United Kingdom these areas of public expenditure have been subject to increasingly severe constraints and in many cases to outright cuts and contractions of employment, in sharp contrast to the experience of the sixties and early seventies. These policies would have to be reversed, especially with respect to education and training, which should be viewed as long-term investment in the future.

Secondly, some services which were provided through the market are now increasingly provided by consumer durables in the household, e.g. washing machines in lieu of laundries and domestic servants (Gershuny, 1983), or television in lieu of cinemas. Employment growth must come in those services which are still income-elastic.

In the remaining service industries, which are predominantly in the private sector, there are an increasing number which have been experiencing rapid technical change and (often for the first time) a sharp increase in capital per person employed. This is the case, for example, with banks, insurance companies, some parts of retail distribution and even some professional services. However, as shown in chapter 10, the expansion of demand and the opening up to 'trade' of services using the new information technology has often been so great as to generate much new employment, together with higher capital intensity. These offer clear new potential for employment and are as critical for the future of IT as machine tools were for engineering in the nineteenth century.

All of this means that problems of technical innovation and the associated structural changes in the economy must be taken into account not only in the primary and secondary sectors, but also now increasingly in the tertiary sector. It remains true that employment in some service sectors is less dominated by the heritage of capital stock and by the labour-saving bias of new capital investment, than is the case with manufacturing and mining. Nevertheless, this distinction is now becoming increasingly blurred and the problem of raising the productivity of capital as well as labour will be increasingly important throughout the economy. We would therefore place great emphasis on technical change in the tertiary sector as one of the main 'engines of growth' for the future and one of the main markets for the electronic equipment produced in the manufacturing sector, as well as the software and IT producer and user services generated in the new rapidly expanding IT service industries themselves.

It is here that our approach diverges most sharply from some mainstream neo-classical and Keynesian approaches. As we have seen, the production function approach is sometimes invoked to justify an emphasis on reduction in real wages in order to 'induce' the adoption of more labour-intensive techniques and to encourage the growth of low wage activities. Small-scale service industries are often cited as among those most likely to benefit from such policies and the abolition of minimum wage constraints and the weakening of trade unions. For example, an article in the *Economist* (28 July 1984, pp. 17–20) by Geldens states:

The demand for work is always there, but not always at earnings which the unemployed have been persuaded to consider as adequate and justified. Shoe-shine boys are nowadays hard to find in Northern Europe, but (at some price) Northern Europeans would prefer to have their shoes cleaned rather than doing it themselves.

The article has many other interesting suggestions on ways to promote higher employment, including retraining for information technology jobs, but the main emphasis is on reduction in both unemployment benefits and real wages. Both this article and other studies have suggested that there is a clear-cut statistical correlation between the level of unemployment and change in real wages, and the high rate of creating new jobs in the US economy over the past 10 years is attributed to the (downward) flexibility of real wages in the United States as compared with Europe. As we have seen, these assumptions are now explicitly incorporated in the Treasury and other econometric models for the United Kingdom.

Whilst not disagreeing with the view that low wage and often part-time, non-union employment can be generated by this route and that this has happened in the United States, we disagree with the view that this is due to substitution of labour for capital, i.e. reversing the main trend of technical change in manufacturing and services. It is rather due to the increased profitability of many manual low technology, labour-intensive service activities.

Whilst some additional employment can be generated in this way and there is certainly scope for many new small firms, we do not see wage reduction as the main route to a stable high employment, high growth economy. As a long-term solution we believe it could lead to a low productivity, low technology, low growth, socially divisive 'shoe-shine boy economy' as in many Third World countries, or to a dual economy, in which one part was high wage and ultra-high technology and the other was based on low wage casual employment. This danger could be aggravated by any tendency to expand the exotic 'Star Wars' type of R & D and production at the expense of civil industries.

The lop-sided concentration of US research and development, massive though this undoubtedly is, on military objectives has contributed to serious structural problems in the bread-and-butter manufacturing industries and services. The United States suffered a serious loss of international competitiveness in many sectors, which was due not only to the prolonged overvaluation of the dollar, but also to a serious and widening technological lag in relation to Japan, and even in some cases to European competitors. The US trade deficit now includes some product groups in which the United States was only recently a technological leader. Both this and the low rate of productivity growth indicate the dangers of the 'dual economy' approach.

The solution which we are advocating is one that uses advanced technology more widely throughout the economy and not just in a few enclaves, and which strives to maintain good employment practices and conditions of work in the services as well as elsewhere. Probably the Swedish economy is nearest to what we have in mind. We would designate it as a 'Technology and work for all' economy. The high technology and IT-producer industries themselves account for only a small proportion of the new jobs which are formed. It is the diffusion to other areas which is crucial, in the first place to the IT-user sector, but ultimately through the entire economy.

Whereas neo-classical growth models tend to make the simplifying assumption of a constant capital–output ratio, our own research reported here and elsewhere (Soete and Freeman, 1985) suggests that the problem of falling productivity of capital has become a serious one in all industrial countries, with the possible exception of Japan, over the past decade or so. This tendency to diminishing marginal productivity of capital has apparently been overcome in the past on the one hand by structural change, i.e. a shift in the pattern of demand towards more labour-intensive sectors, with lower than average investment needs per unit of output, and on the other hand by major technical changes in various sectors of the economy, which brought about a rise in capital productivity, reversing the previous downward trend.

We certainly would not underestimate the importance of the remaining labour-intensive areas of employment in advanced economies. On the contrary we would stress that they are the main sources of future employment growth and that there are certain types of service which are in principle rather labour intensive, in the sense that direct human contact is the very essence of the service or activity. These include many types of caring and personal services, such as childcare, psychiatric care, counselling services, many types of health care, and (we would say) much education and training. They also include many types of creative work and leisure activities including artistic, scientific and craft activities. Typically there is no satisfactory way of measuring 'productivity' in such activities but this does not detract from their importance.

The provision of these services has depended historically on a combination of growing public provision (usually on a non-profit basis) and voluntary part-time provision, as well as the private market mechanism. The growth of such services is, we would maintain, one of the hallmarks of a civilized society, but for that growth to be sustained, and employment growth to continue in this area, then it is essential to sustain high rates of productivity increase in the rest of the economy. This points to the great importance of reversing the downward trend in the marginal productivity of capital in the market sector of the economy. Instead of moving backwards along existing production functions, the move should be to new production functions with lower labour and capital costs per unit of output.

We have argued that the impetus to a major upswing of the economy could come in principle, as it has come in the past, from the adoption of a major new technological paradigm. Such a transformation could create the basis for renewed advances in capital productivity and in the use of energy and materials, as well as in labour productivity. Thus, whereas more orthodox economic policies would put the main emphasis on the relative price of labour and on the substitution of labour for capital, we would emphasize the new technologies and their capacity for generating expansion in the economy generally and public policies directed to the investment, education, and training needed to promote their widespread adoption.

In relation to future employment trends, we have stressed the particular significance of capital costs, because of the importance of generating new employment quickly with relatively small increments of new investment. Where this is not possible, and/or when the scale of investment is very large and of an infrastructural type (as was the case, for example, with electric power and the motorways in the past) there is indeed a case for big programmes of public investment. This is particularly important when there is a 'technological multiplier', i.e. when the secondary effects are not simply of the conventional type of Keynesian public works, but promote the widespread adoption of information revolution to many types of communication system as well as to education and training. It is this 'rationale' which underlies the ambitious French and Japanese policies. There could be a major role for public investment in the 'cabling' infrastructure necessary for the full development of many types of information service.

However, the overall growth and employment stimulus from such public investment would be far more significant, and less constrained by both budget-deficit and inflationary considerations, if there are substantial capital-saving gains to be made from such an investment. Both computers and telecommunications equipment now appear to offer substantial potential for capital saving. IT appears to 'embody' much potential pervasive productivity gain. For such gains in capital productivity to generate widespread effects throughout the economy, it will however be necessary to make significant advances in the design and development in other types of capital goods, such as robots, sensors, process control instrumentation, and so forth. There is some evidence of parallel gains in these areas too, but for the potential benefits to be realized, much more technical change will be needed in sectors far removed from the electronic and communication industries.

Whether or not such capital and labour productivity gains can be realized throughout the economy depends in the final instance, though, upon whether the type of socio-institutional problems emphasized by Perez (1985) can be resolved. If changes in the institutional framework can lead to a good 'match' with the characteristics of the new technological paradigm, then the potential gains could be realized and a new wave of economic expansion would be feasible.

It is however essential not to underestimate the scope of institutional change which is needed. As we have indicated in chapter 12, it will involve big changes in the educational and training systems; in management and labour attitudes; the pattern of industrial relations and worker participation; in working arrangements; in the pattern of consumer demand; in the conceptual framework of economists, accountants and governments; and in social, political and legislative priorities.

Perhaps the most important point is that 'intangible' capital investment must now be recognized in its own right as more important than the transitory physical capital investment, which is today still the main focus of attention for most managements, accountants and economists. For a long time firms in the computer industry (and other R & D intensive industries) have already devoted greater resources to R & D, education and training, information services, design and software development, than to physical capital investment. This balance will now be tilted even more towards intangible investment as the information system available to firms, government departments and other institutions is becoming its most critical resource. There is of course a very close link between the 'intangible' software, etc., and the 'tangible' hardware in an information system, but it is increasingly necessary for the 'intangible' resources to be recognized fully as the main focus for strategic long-term development. This means that they must be considered as a form of investment and not as consumption or current expenditure.

This applies both to formal information systems and data banks and to the costs of developing, educating and training the people involved. The scale of the change in skills and in occupations is still underestimated. Even in a period of high unemployment there are still persistent skill shortages for certain types of labour, particularly in relation to electronic engineering, software designs, and systems analysis. In addition, there are many types of skilled people whose level of training and qualification is inadequate for the new types of work which they are being called upon to perform, or ought to be able to perform. This applies perhaps most of all to management at all levels, but it also applies to many types of craftsmen, for example maintenance workers, who frequently lack the requisite combination of electromechanical with electronic skills.

The information revolution also affects the whole climate and conduct of industrial relations. In principle it would be possible to introduce computerization in the style of '1984'. It does make possible the most sophisticated centralized type of 'Big Brother' time and motion study, whether of sales workers in a supermarket, or of miners operating an underground shearer-loader, but it also makes possible a very high degree of decentralization, local responsibility and initiative if the computerization and information systems are designed in that way, and if industrial relations are properly conducted.

Computerization could also facilitate the introduction of much greater flexibility in working hours. In many occupations, flexitime has already been introduced, and the scope of this will be increased. Part-time work and work-sharing are likely to become more widespread. But here again this social change can take a variety of different forms. It could be introduced as a means of depressing wage levels and of reducing social insurance benefit, if part-time workers, and especially female workers, are treated as a lower grade type of labour. Or it could be introduced mainly as a response to the desire of many workers, male as well as female, at various stages in their working lives to work part-time, so that they can spend more time with young children, on pursuing education and training, or other activities. It ought to be possible to humanize working arrangements in this way, but it will not be easy in view of the enormous strength of the old management and union attitudes. The involvement of the social sciences and humanities in developing social policies is just as critical as the technology itself.

The institutional framework for future economic recovery is only now being shaped. Most of our institutions and ideologies are still geared to the old postwar technological paradigm. Only through social and political debate and conflict shall we determine how we reshape our institutions and our way of life to match the potential of the new technology and to humanize its innumerable potential applications. The new patterns of employment which will emerge should be of a kind which encourage great variety in hours of work and in continuing education and training, but which ensure to everyone who is seeking paid employment, the opportunities to work in socially useful activity. This should mean a renewed commitment by society to the goal of 'full employment', but in a new social context which takes account both of changes in technology and of changes in society.

References

Allen, G. C. (1981): Industrial Policy and Innovation in Japan. In: C. Carter (ed.) *Industrial Policy and Innovation*. London: Heinemann.

Alvey Committee (1982): A programme for advanced information technology. London, HMSO.

Ando, A. K., Modigliani, F., Rasche, R. and Turnovsky, S. J. (1974): On the role of expectations of price and technological change in an investment function. *International Economic Review*, 15.

Armstrong, A. G. (1976): Capital stock in UK manufacturing: disaggregated estimates 1947–76. In: K. D. Patterson and K. G. Schott (eds) *The Measurement of Capital*. London: Macmillan.

Arnold, E. and Guy, K. (1986): *Parallel Convergence: National Strategies in Information Technology*. London: Frances Pinter.

Arrow, K. (1962): Economic welfare and the allocation of resources for inventions. In: (NBER) *The Rate and Direction of Inventive Activity*. NBER Princeton University.

Arthur, B. (1985): Competing technologies and lock-in by historical small events. The dynamics of allocation under increasing returns. In: *CEPR Working Paper*. Stanford University Press.

Arthus, B. (1983): Capital, energy and labour substitution: the supply block in OECD medium-term models. Working Paper No. 2, Paris: OECD.

ASTMS, (1978): *Technological change and Collective Bargaining*. London: ASTMS.

Atkinson, A. and Stiglitz, J. (1969): A new view of technological change. *Economic Journal*, 79, 573–78.

Bagrit, L. (1965): *The Age of Automation*. (BBC Reith Lectures 1964), London: Weidenfeld and Nicolson.

Ball, R. J., Burns, T. and Warburton, P. J. (1979): The London business school model of the UK economy. In: J. C. Ormerid (ed.) *Economic Modelling*. London: Heinemann.

Bank of England (1985), Services in the UK Economy. *Bank of England Quarterly Bulletin*, September.

Barker, T. S. (ed.) (1976): *Economic Structure and Policy*. London: Chapman and Hall.

Barker, T., Borooah, V., van der Ploeg, R. and Winters, A. (1980): The Cambridge multisectoral dynamic model: an instrument for national economic policy analysis. *Journal of Policy Modelling*, 3, pp. 319–344.

Barras, R. (1983): *Growth and Technical Change in the UK Service Sector*. London: Technical Change Centre, February.

Barron, I. and Curnow, R. (1979): *The Future with Micro-electronics*. London: Frances Pinter.

Begg, D. K. H. (1982): *The Rational Expectations Revolution in Macroeconomics*. Oxford: Philip Allan.

Bell, D. (1974): *The Coming of Post-Industrial Society*. London: Heinemann.

Bide Committee (1986): *Information Technology, the National Resource, a Plan for Concerted Action*. London: HMSO.

Bird, J. (1985): IBM suppliers get first public warning. *Sunday Times*, May 19.

Bourniatin, M. (1933): Technical progress and unemployment. *International Labour Review*, 273, pp. 327–348.

Brady, T. and Senker, P. (1986): *Contract Maintenance: No Panacea For Skill Shortages*. Manpower Services Commission.

Breakwell, Collice, G.M.A., Hanson, B. and Propper, C. (1984): Attitudes towards the unemployed, effects of theoretical identity. *British Journal of Social Psychology*, 23, 87–8.

Brenner, M. H. (1976): *Estimating the Social Costs of National Economic Policy Implications for Research and Physical Health and Criminal Aggression*. Joint Economic Committee of Congress, Paper No. 5, Washington DC: US Government Printing Office.

Brittan, S. (1985): Economic Viewpoint. In *Financial Times*, 17 October.

Brodie, I. (1986): Distribution traders. In: A. D. Smith (ed.) *Technological trends and employment, commercial service industries*, Aldershot: Gower Publications.

Browning, H. C. and Singelman, J. (1978): The transformation of the US labour force: the interaction of industry and occupation. *Politics and Society*, 8, 481–509.

Bruno, M. and Sachs, H. (1985): *Economics of Worldwide Stagflation*, Cambridge: Harvard University Press.

Buiter, W. H. (1984): Allocative and stabilisation aspects of budgetary and financial policy. Inaugural Lecture at LSE, Centre for Economic Policy Research, London, November.

City University Business School, (1984): *Economic Review*, Spring.

Clark, C. A. (1940): *The Conditions of Economic Progress*. London: Macmillan.

Clark, G. (1984): *Structural Unemployment, Efficiency and Innovations: English Agriculture in the 19th Century*. Stanford University Press.

Clark, J. (1983): Employment projections and structural change. In: D. L. Bosworth (ed.) *The Employment Consequences of Technical Change*. London: Macmillan.

Clark, J. (ed.) (1985): *Technological Trends and Employment: Basic Process Industries*. Aldershot: Gower Publications.

Cooper, C. M. and Clark, J. A. (1982): *Employment, Economics and Technology: The Impact of Technical Change on the Labour Market*. Brighton: Wheatsheaf Books Ltd.

David, P. (1975): *Technical Choice, Innovation and Economic Growth*. Cambridge: Cambridge University Press.

David, P. (1986): New technology diffusion, public policy and international competitiveness. In: R. Landau and N. Rosenberg (eds) *The Positive Sum Strategy: Harnessing Technology for Economic Growth*. Washington: NAS, National Academy Press.

de la Mothe, J. R. (1986): Financial services. In: A. D. Smith (ed.) *Technological Trends and Employment: Commercial Service Industries*. Aldershot: Gower Publications.

Devine, W. (1983): 'From shafts to wires: historical perspectives' *Journal of Economic History*, 43, pp. 347–373.

Dosi, G. (1982): Technological paradigms and technological trajectories. *Research Policy*, 11, 147–63.

Dosi, G. (1984): *Technical Change and Industrial Transformation*. London: Macmillan.

Dosi, G. and Soete, L. (1983): Technology gaps and cost-based adjustments: some explorations on the determinants of international competitiveness. *Metroeconomica*, XXXV, Bologna.

Douglas, P. (1930): Technological unemployment. *American Federationist*. 37 (8), pp. 923–950.

Driehuis, W. (1979): Capital labour substitution and other potential determinants of structural employment and unemployment. In: Paris: OECD *Structural Determinants of Employment and Unemployment*.

Einzig, P. (1957): *The Economic Consequences of Automation*. London: Secker and Warburg.

Eisenberg, P. and Lazarsfeld, P. F. (1938): The psychological effects of unemployment. *Psychological Bulletin*, 35, 358–90.

Englander, P. (1985): The economics of innovation in telecommunications. PhD thesis, London Graduate School of Business Studies.

Etzione, A. and Jargorsky, P. (1984): High tech basic industry and the future of the American economy. *Human Resources Management*, 23, 229–40.

Fano, E. (1984): The problem of 'technological unemployment' in the industrial research of the 1930's in the United States. *History and Technology*, 1, 277–306.

Fisher, A. G. B. (1939): *The Clash of Progress and Security*. London: Macmillan.

Forty, A. (1981): *Objects of Desire*. London: Thames & Hudson.

Freeman, C. (1978): Technology and unemployment: long waves in technical change and economic development. Holst Memorial Lecture, Eindhaven.

Freeman, C. (1979): Technical change and unemployment. In: S. Encel and J. Ronayne (eds) *Science, Technology and Public Policy*. Oxford: Pergamon Press.

Freeman, C. (1982): Some economic implications of microelectronics. In: C. D. Cohen (ed.) *Agenda for Britain: micro policy choices for the 80's*, Oxford: Philip Allan, pp. 53–88.

Freeman, C. (1985): *Engineering and Vehicles*. Aldershot: Gower Publications.

Freeman, C. and Soete, L. (1985): *Information Technology and Employment: An Assessment*. Brussels: IBM.

Freeman, C., Clark, J. A. and Soete, L. (1982): *Unemployment and Technical Innovation: A Study of Long Waves and Economic Development*. London: Frances Pinter.

268   References

Friedman, M. (1968): The role of monetary policy. *American Economic Review*, 58, 1–17.

Fuchs, V. F. (1968): *The Service Economy*. New York: National Bureau of Economic Research.

Fuchs, V. (ed.) (1969): *Production and Productivity in the Service Industries*. New York: NBER.

Furnham, A. (1982): Explanations of unemployment in Britain. *European Journal of Social Psychology*, 12, 335–52.

Geldens, A. (1984): *Economist*, 25 July, pp. 17–20.

Gershuny, J. (1978): *After Industrial Society*. London: Macmillan.

Gershuny, K. and Miles, I. (1983): *The New Service Economy/The Transformation of Employment in Industrial Societies*. New York: Praeger Publishers.

Gooding, K. (1984): The Cultural Revolution at GM. *Financial Times*, November 20.

Gourvitch, A. (1940): *Survey of Economic Theory on Technological Change and Employment*. New York: Augusta M. Kelley.

Gregory, T. (1931): Rationalization and technological unemployment. *Economic Journal*, 41, 551–67.

Griffin, T. (1976): The stack of fixed assets in the United Kingdom. *Economic Trends*, October.

Guy, K. (ed.) (1984): *Basic Consumer Goods*. Aldershot: Gower Publications.

Guy, K. (1986): Policies in the UK Electronics and Information Technology Sector. SPRU Working Paper, March.

Hahn, F. and Matthews, R. (1967): The theory of economic growth. *Surveys of Economic Theory*, II, 1–124.

Hansen, A. (1932a): Institutional features and technological unemployment. *Quarterly Journal of Economics*, XLV, 684–97.

Hansen, A. (1932b): The theory of technological progress and dislocation of employment. *American Economic Review*. XXII (1), pp. 25–31.

Harrod, R. (1948): *Towards a Dynamic Economics*. London: Macmillan.

Heckscher, E. (1935): *Mercantilism*. London, Allen & Unwin.

Hicks, J. (1935): *The Theory of Wages*. London: Macmillan.

Holden, K. et al. (1982): *Modelling the UK Economy*. Oxford: Martin Robertson.

Hollander, S. G. (1965): *The Sources of Increased Efficiency: A Study of DuPont Rayon Plants*, MIT Press.

House of Lords: (1985): *Report from the Select Committee on Overseas Trade*. London: HMSO.

Information Technology Advisory Panel (1983): *Making a Business of Information: A Survey of New Opportunities*. London: HMSO.

Jahoda, M. (1982): *Employment and Unemployment: A social-psychological analysis*. Cambridge: Cambridge University Press.

Jerome, H. (1934): *Mechanization in Industry*. New York: NBER.

Jones, D. I. H. (1980): Technological Change, Demand and Employment. University of Liverpool, School of Economics Studies Discussion Paper series, No. 101, mimeo.

Jones, D. (1985): Vehicles. In: C. Freeman (ed.) *Technological Trends and Employment*, 4 (21). Aldershot: Gower Publications.

Kaldor, N. (1932): A case against technological progress. *Economica*, XII, 180–96.

Katsoulacos, Y. (1984): Product innovation and employment. *European Economic Review*, 26, 83–108.

Katsoulacos, Y. (1986): *The Employment Effect of Technical Change: A Theoretical Study of New Technology and the Labour Market*. Brighton: Wheatsheaf Books Ltd.

Katouzian, M. (1970). The development of the service sector: A new approach. *Oxford Economic Papers*, 22, 362–82.

Kay, J. and Morris, N. (1982): No longer rich on the dole? *New Society*, 22, 267–268.

Keirstead, B. S. (1948): *The Theory of Economic Change*. Toronto: Macmillan.

Kennedy, C. (1964): Induced bias in innovation and the theory of distribution. *Economic Journal*, 75, 541–7.

Kennedy, C. and Thirwall, A. (1973): Technical progression (ed.) Royal Economic Society. *Survey of Applied Economics*, 1, 115–77.

Keynes, J. M. (1930): *Treatise on Money*, 2. London: Macmillan.

Keynes, J. M. (1936): *The General Theory of Employment, Interest and Money*. New York: Harcourt, Brace & Co.

Kimbel, D. (1985): Microelectronics and employment. In: *International Symposium on Microelectronics and Labour*. National Institute of Employment and Vocational Research: Tokyo, September.

Kindleberger, C. F. (1985): *Keynesian v Monetarism*. London: Allen and Unwin.

Klamer, A. (1984): *The New Classical Macroeconomics*. Brighton: Wheatsheaf Books Ltd.

Kline, S. and Rosenberg, N. (1985): An overview of the process of innovation. In: R. Landau and N. Rosenberg (eds) *The Positive Sum Strategy: Harnessing Technology for Economic Growth*. Washington: NAS, National Academy Press.

Kuhn, T. (1962): *The Structure of Scientific Revolutions*. Chicago: University Press.

Kuwahara, Y. (1984): Creating new jobs in high technology industries, paper presented at OECD inter-governmental conference on employment growth in the context of structural change. Paris: OECD.

Kuznets, S. (1972): *Modern Economic Growth*. New Haven: York University Press.

Layard, R. and Nickell, S. (1985): The causes of British unemployment. *National Institute Economic Review*, 113.

Lederer, E. (1938): *Technischer Fortschritt und Arbeitslosigkeit*, Tubingen: J. C. B. Mohr, 1911, also translated as *Technical Progress and Unemployment*. Geneva: International Labour Office.

Lindley, R. M. (ed.) 1980: *Economic Change and Employment Policy*. London: Macmillan.

Machlup, F. (1962): *The Production and Distribution of Knowledge in the United States*. Princeton University Press.

Maclean, J. M. and Rush, H. J. (1978): The impact of micro-electronics in the U.K.: a suggested classification and illustrative case studies. SPRU Occasional Paper Series, University of Sussex, No. 7, June.

Mackintosh, I. (1986): *Sunrise Europe: The Dynamics of Information Technology*, London: Basil Blackwell.

Mansell, R. (1986): A background paper on the tradeable information sector: research issues and priorities, 25 April, mimeo.

Marquand, J. (1978): *The Service Sector and Regional Policy in the United Kingdom*. London Centre for Environmental Studies.

Matthews, K. (1984): Unemployment between the wars. *University of Liverpool Quarterly Economic Bulletin*, May.

Mensch, G. (1975): *Das Technologische Patt: Innovationen uberwinden die Depression*, Frankfurt: Umschau, English edition (1979): *Stalemate in Technology: Innovations Overcome the Depression*. New York: Ballinger.

Micklewright, J. (1985): Male unemployment and the family expenditure survey 1972–80. *Oxford Bulletin of Economics & Statistics*, 46 (1), Feb. 1984.

Miles, I. (1983): Adaptation to unemployment, Occasional paper No. 20, Sussex University Press, SPRU.

Mills, F. (1932): *Economic Tendencies in the United States*. New York: NBER.

Minford, P. (1984): *Unemployment – Cause and Cure*. Oxford: Martin Robertson.

Momigliano, F. and Siniscalco, D. (1983): The growth of service employment: a reappraisal. *Banca Nazionale del Laboro Quarterly Review*, September, pp. 269–306.

Neary, P. (1980): On the short-run effects of technological progress. *Oxford Economic Papers*, 32, 224–34.

Neisser, H. (1942): Permanent technological unemployment. *American Economic Review*, 32 (1), 50–71.

Nelson, R. and Winter, S. (1977): In search of a useful theory of innovation. *Research Policy*, 6, 36–76.

Nelson, R. and Winter, S. (1982): *An Evolutionary Theory of Economic Change*. Cambridge Massachusetts: Harvard University Press.

Nickell, S. J. (1978): *The Investment Decisions of Firms*. Cambridge: Nisbet.

Nickell, S. J. (1984): A Review of *Unemployment: Cause and Cure*, by Patrick Milford with David Davies, Michael Peel and Alison Speaque. *Economic Journal*, 94, 946–53.

NIESR (1983): *Economic Review*, November, No. 107.

NIESR (1984): *Economic Review*, November, No. 111.

Northcott, J. (1986): *Microelectronics in Industry. Promise and Performance*. London: Policy Studies Institute.

Northcott, J. and Rogers, P. (1982): *Microelectronics in Industry: What's Happening in Britain*. London: Policy Studies Institute.

Northcott, J. and Rogers, P. (1985): *Microelectronics in British Industry: the Pattern of Change*. London: Policy Studies Institute.

OECD (1977): *Towards Full Employment and Price Stability*. Paris: McCracken Report.

OECD (1979): *Determinants of Employment and Unemployment*. Vol. 2, Paris: OECD.

OECD (1981): *Information Activities, Electronics and Telecommunications Technologies: Impact on Employment, Growth and Trade*. Vol. 1, Paris: OECD.

OECD/ICCP (1982): *Microelectronics, Robotics and Jobs*. Paris: Computer & Communications Committee for Information.

OECD (1983): *The Present Unemployment Problem*. Paris: Working Party No. 1 of the Economic Policy Committee.

OECD (1985): *Employment Outlook*. Paris: OECD.

Olsson, M. (1982): *The Rise and Decline of Nations*. New Haven: Yale University Press.

Office of Technology Assessment (1985): *Information Technology R & D*. Washington D.C: OTA.

Panic, M. (1978): *Capacity Utilization in UK Manufacturing Industry*, London: National Economic Development Office, Discussion Paper 5.

Pasinetti, L. (1981): *Structural Change and Economic Growth: A Theoretical Essay on the Dynamics of the World of Nations*. Cambridge: Cambridge University Press.

Patel, P. and Soete, L. (1984): Information technologies and the rate and direction of technical change. *Proceedings of the OECD Expert Meeting on Information Technology and Economic Perspectives*. DSTI/ICCP/84.20, Paris: OECD.

Pavitt, K. (ed.) (1980): *Technical Innovation and British Economic Performance*. London: Macmillan.

Pavitt, K. (1983): High technology. In: G. D. Cohen (ed.) *The Common Market: 10 Years After*. London: Philip Allan.

Pavitt, K. (1984): Sectoral patterns of technical change: towards a taxonomy and a theory. *Research Policy*, 13 (6), 343–73.

Pavitt, K. (1985): Patent statistics as indicators of innovation activities. *Scientometrics*, 7, 77–99.

Pavitt, K., Robson, M. and Townsend, J. (1985a & b): *The Size Distribution of Innovating Firms in the UK 1945–1983*. DRC Discussion Paper No. 25, SPRU, University of Sussex.

Perez, C. (1983): Structural change and assimilation of new technologies in the economic and social systems. *Futures*, 15, 357–75.

Perez, C. (1985): Micro-electronics, long waves and world structural change: new perspectives of developing countries. *World Development*, 17, 441–63.

Persnick, V. (1984): The job outlook. *Monthly Labour Review*, November, 30–32.

Petty, W. (1899): Verbum sapienti. In: C. J. Hull (ed.) *Economic Writings*. Cambridge: Cambridge University Press.

Piore, M. J. and Sabel, C. F. (1984): *The Second Industrial Divide: Possibilities for Prosperity*, New York: Basic Books Inc.

Porat, M. (1976): *The Information Economy*, Michigan: Ann Arbor.

Quarterly Industrial Trends Survey (1985): conducted by the Confederation of British Industry.

Quinn, J. (1986): The impacts of technology in the services sector, paper prepared for a Symposium on World Technologies and National Sovereignty, NAE Washington, 13–16 February.

Ricardo, D. (1821): Principles of political economy and taxation. In: P. Sraffa (ed.) *The Works and Correspondence of David Ricardo*. 3rd edn, Cambridge: Cambridge University Press.

Roach, S. (1986): Macro realities of the information economy. In: R. Landau and N. Rosenberg (eds) *The Positive Sum Strategy: Harnessing Technology for Economic Growth*. Washington: NAS, National Academy Press, Washington.

Robertson, D. (1931): The world slump. In: A. Pigou and D. Robertson, *Economic Essays and Addresses*. London: P. King & Son.

272   References

Robertson, J., Briggs, J. and Goodchild, A. (1982): *Structure and Employment Prospects of the Service Industries*. Research Paper No. 30, Department of Employment, July.

Rosenberg, N. (1976): *Perspectives on Technology*. Cambridge: Cambridge University Press.

Rosenberg, N. (1982): *Inside the Black Box: Technology and Economics*. Cambridge: Cambridge University Press.

Rothwell, R. and Zegveld, W. (1979): *Technical Change and Employment*. London: Frances Pinter.

Rush. H. and Soete, L. (1984): Clothing. In: K. Guy (ed.) *Technological Trends and Employment: Basic Consumer Goods*. Aldershot: Gower Publications.

Salter, W. (1961): *Productivity and Technical Change*. Cambridge: Cambridge University Press.

Schonberger, D. (1982): *Japanese Manufacturing Technologies: Nine Lessons in Simplicity*. New York: Macmillan.

Schumpeter, J. (1939): *Business Cycles: A Theoretical, Historical and Statistical Analysis of the Capitalist Process*. 1st edn, New York: McGraw-Hill.

Senker, P. and Arnold, E. (1982): *Designing the Future: the Implications of CAD Interaction Groups for Employment Skills*. EITB, Paper No. 9.

Senker, P. and Brady, T. (1986): *Contract Maintenance: No Panacea For Skill Shortages*. Manpower Services Commission.

Sheffrin, S. M. (1984): *Rational Expectations*. Cambridge: Cambridge University Press.

Sinclair, P. J. N. (1981): When will technical progress destroy jobs? *Oxford Economic papers*, 33, 1–18.

Singelmann, J. (1979): *From Agriculture to Services*. Beverley Hills: Sage Publications.

Singelmann, J. (1978): The sectoral transformation of the labour force in seven industrial countries 1920–70. *American Journal of Sociology*. Vol. 83, No. 5.

Smith, A. D. (1972): *The Measurement and Interpretation of Service Output Changes*. London: National Economic Development Office.

Smith, A. A. D. (1986): *Commerical Service Industries: Technological Trends and Employment*, Vol. 5. Aldershot: Gower Publications.

Soete, L. (1981): *A general test of technological gap trade theory, Weltwirtschaftliches Archiv*, 4, 638–59.

Soete, L. and Dosi, G. (1983): *Technology and Employment in the Electronics Industry*. London: Frances Pinter.

Soete, L. and Freeman, C. (1985): New technologies investment and employment growth in OECD. *Employment Growth and Structural Change*, Paris: OECD.

Soete, L. and Patel, P. (1985): Recherche et Developpement, Importations de Technologie et Croissance Economique: Une Analyse Internationale. *Revue Economique*. Vol. 36, No. 5, September 1985, 975–1000.

Soete, L. and Turner, R. (1984): Technology diffusion and the rate of technical change. *The Economic Journal*, 94, 612–23.

Soete, L. (1985a): Innovatie en Werkgelegenheid. In: 17de VWEC, *Innoveren en Ondernemen*, Antwerp.

Soete, L. (ed.) (1985b): Electronics and Communications. *Technological Trends and Employment*. Aldershot: Gower Publications.

Solow, R. (1957): Technical change and the aggregate production function. *Review of Economics and Statistics*, 39, 312–20.

Stiglitz, J. (1985): Equilibrium wage distributions, *Economic Journal*, 95, pp. 595–618.

Stoneman, P. (1983): *The Economic Analysis of Technological Change*. Oxford: Oxford University Press.

Stoneman, P. (1984): An analytical framework for an economic perspective on the impact of new information technologies. *ITEP-project*. Paris: OECD.

Surrey, J. and Thomas, S. (1980): *Worldwide Nuclear Plant Performance*. SPRU Occasional Paper No. 10, University of Sussex.

Townsend, J. (1976): *Innovations in Coal Mining Machinery*. SPRU Occasional Paper No. 3, University of Sussex.

US National Commission on Technology, Automation and Economic Progress (1966): *Technology and the American Economy*, Report to the US Congress, 6 Volumes, Washington.

US Industrial Outlook, (1985): Department of Commerce.

Venables, A. (1985): The employment effect of product and process innovation. *Oxford Economic Papers*, 37, pp. 230–248.

von Mises, L. (1936): *Socialism*. London: Jonathan Cape.

Walker, W. (1985): Information technology and the use of energy. *Energy Policy*, Vol. 14, No. 6, pp. 466–488.

Warr, P. (1983): Work, jobs and unemployment. *Bulletin of the British Psychological Society*, 36, 305–34.

Weintraub, D. (1937): Unemployment and increasing productivity. In: National resources committee *Technological Trends and National Policy*, Washington D.C.

von Weizsacker, C. 1966: Tentative notes on a two sector model with induced technical progress. *Review of Economic Studies*, 245–51.

Wigley, K. J. (1970): Production models and time trends of input-output coefficients. In: W. F. Gossling (ed.) *Input-Output In The United Kingdom*. Cass.

Wyatt, G. (1986): *The Economics of Invention: A Study of the Determinants of Inventive Activity*. Brighton: Wheatsheaf Books Ltd.

The Gower
Sector Studies

These were published by Gower Press during 1984, 1985 and 1986 and are referred to in this book as 'The Gower Series'. The Sector Studies in each volume, together with their authors, are listed below:

Volume 1 Basic Consumer Goods edited by K. Guy 1984

Footwear	K. Guy
Food, Drink and Tobacco	J. Clark
Textiles	L. Soete
Clothing	H. Rush and L. Soete

Volume 2 Basic Process Industries edited by J. Clark 1984

Energy	G. F. Ray
Chemicals	J. Clark
Paper	K. Guy
Iron and Steel	J. Aylen and J. Clark

Volume 3 Electronics and Communications edited by L. Soete 1985

Electronics	L. Soete
Communications	K. Guy
Instrumentation	J. Rendeiro
Printing	K. Guy and W. Haywood

Volume 4 Vehicles and Engineering edited by C. Freeman 1985

Mechanical Engineering	C. Freeman
Electrical Engineering	C. Freeman
Metal Goods	C. Freeman
Vehicles	D. Jones

Volume 5 Commercial Service Industries edited by A. D. Smith 1985

Miscellaneous Services	A. D. Smith
Financial Services	J. R. de la Mothe
Distributive Trades	I. Brodie

Volume 6 Other Industries (forthcoming) edited by P. Patel 1986

Construction	D. Gann
Building Materials	D. Gann
Transport	I. Brodie and P. Patel

Index

Indexed by Elizabeth Clutton